THE HOMEOWNER'S HANDBOOK OF
SOLAR WATER HEATING SYSTEMS

THE HOMEOWNER'S HANDBOOK OF
SOLAR WATER HEATING SYSTEMS

How to Build or Buy Systems to Heat Your Water, Your Swimming Pool, Hot Tub or Spa

BY BILL KEISLING

 Rodale Press, Emmaus, Pa.

Printed in the United States of America on recycled paper containing a high percentage of de-inked fiber.

Book design by Jerry O'Brien
Illustrations by Frank Rohrbach, except:
   Figure 2-1 (p. 18), reprinted by permission of Cheshire Books from *A Golden Thread*, © 1980 by Ken Butti and John Perlin.
   Figures 3-3 through 3-10 (pp. 42–49) adapted from *The Passive Solar Energy Book*, © 1979 by Edward Mazria, with permission of Rodale Press, Emmaus, Pa.
   Figure 3-11 (p. 50) reprinted from *Solarizing Your Present Home*, Joe Carter, ed., © 1981 Rodale Press, Emmaus, Pa.
Photography by Rodale Press Photography Department, except:
   Photo 1–9 (p. 12), courtesy of International Technology Sales Corp., Englewood, Colo.
   Photo 2–1 (p. 17) and photo 5–1 (p. 117), reprinted by permission of Cheshire Books from *A Golden Thread*, © 1980 by Ken Butti and John Perlin.
   Photo 2–2 (p. 22), courtesy of Solar Kinetics, Inc., Dallas, Tex.
   Photo 2–3 and photo 2–4 (p. 23), courtesy of Ken Butti.
   Photo 2–6 (p. 27) and photo 6–8 (p. 162), courtesy of Chris Fried Solar, Catawissa, Pa.
   Photo 6–4 (p. 145), courtesy of George A. Vorsheim.
   Photo 6–11 (p. 173), courtesy of American Solar King Corp., Waco, Tex.
   Photo 7–4 (p. 180), courtesy of Holly Solar Products, Inc., Petaluma, Calif.
   Photo 8–2 and photo 8–3 (p. 193), courtesy of Cover Pools, Inc., Murray, Utah.

The collectors shown on the cover were manufactured by Bio-Energy Systems, Inc., Ellenville, N.Y., and supplied by NEWAYS, Inc., Ridgefield, Conn.

Library of Congress Cataloging in Publication Data

Keisling, Bill.
   The homeowner's handbook of solar water heating systems.

   Bibliography: p.
   Includes index.
   1. Solar water heating. 2. Solar heating. I. Title.
TH6561.7.K44   1983   696'.6   82-25083
ISBN   0-87857-444-1 hardcover
ISBN   0-87857-445-X paperback
2 4 6 8 10 9 7 5 3 1 hardcover
2 4 6 8 10 9 7 5 3 1 paperback

CONTENTS

ACKNOWLEDGMENTS

Thanks to the farsightedness of Bob Rodale, solar energy research has been conducted at Rodale Press for many years; I am indebted to the solar energy researchers at Rodale Press who tested and improved many of the solar water heating systems described in this book. Bob Flower helped whenever I needed technical assistance. I also received invaluable information and assistance from Margaret J. Balitas, Jim Eldon, Fred Langa, Roger Moyer, Kip Perkins and Harry Wohlbach. Special thanks to Joe Carter, who edited this book.

INTRODUCTION

For the last several years I've had a keen interest in America's energy policies. This dates back to a cold morning in March, 1979, when I earned the dubious distinction of being the first newspaper reporter on the scene of the accident at the Three Mile Island nuclear power plant. Before this, energy was something I took for granted or, at most, groused about when I bought gasoline. I knew energy was becoming expensive, but only after seeing that my family, friends and hometown could have been blown off the face of the earth by a nuclear meltdown did I begin to realize that the cost of America's energy policy was becoming catastrophically dear.

Why, then, would I write a book about solar water heating? Of the 76.2 quadrillion Btu (quads) of energy consumed by Americans in 1980, just 2.7 quads, about 3.5 percent, of America's energy needs were supplied by nuclear power. About 4 percent of our energy consumption is spent on residential and commercial water heating. What if every home and business in America had solar-heated water? There would be no need for nuclear power.

Common sense dictates another conclusion about the way water is heated by electricity. In a nuclear, coal- or oil-fired power plant, water is heated to several thousand degrees Fahrenheit to make steam, which turns a turbine, which generates electricity, which is sent over wires to your home, where it powers your water heater to give you 120° to 140°F water. It turns out that it takes three units of energy at the same power plant to get one unit into your water heater. Yet the sun has no trouble heating water from 50° to 120°F; much less heat is wasted in a household solar water heating system, and in most cases the cost of solar energy is less for homeowners than the cost of conventional fuels.

Solar heating is a better idea, and it should be talked out. This book is backed by more than seven years of research and development at Rodale Press. Over the years several systems were designed, built, monitored and refined by Rodale engineers, and a great deal of information was gathered. My aim has been to report this information so it can be understandable to consumers, and my hope is that you'll find this book a useful answer to your interest in solar water heating.

Chapter 1 explains how you can conserve hot water to make a considerable cut in your present water heating costs and to help reduce the cost of any solar water heating system. Chapter 2 introduces the systems that put the sun into hot water. Chapter 3, "Choosing the Right System," helps you do just that by presenting a simple decision-making process. You'll also be able to calculate the return on your solar investment. Instructions for building a batch collector are found in chapter 4, and chapter 5 explains everything you need to know to build a flat plate collector. Chapter 6 shows how to build or buy the systems that are used with flat plates. Coal and woodstove water heating systems are discussed in chapter 7, along with the latest information on heat pump water heaters. Solar heating for swimming pools, hot tubs and spas is the topic of chapter 8, and chapter 9 presents advice for finding and working with a solar contractor.

As you'll see in these chapters, solar water heating makes sense technically and economically. The same can't always be said about other of this country's energy systems.

Following the accident on Three Mile Island the ratepayers of the Metropolitan Edison Company were asked to pay the bulk of a billion-dollar cleanup bill. The utility proposes to spend another $100 million or so refurbishing the so-called undamaged Unit One reactor, which was shut down since before the 1979 accident and has seriously degraded. One billion, one hundred million dollars. That much money could buy a solar water heater for every one of Met-Ed's 327,000 residential ratepayers, leaving the rest for cleanup and conversion of Three Mile Island into a monument dedicated to common sense. With conservation and solar energy we can put more sense into this country's energy policies.

1

CONSERVING HOT WATER

Before investing in solar water heating equipment you'll want to make your existing hot water system more efficient. As is usually the case in any energy retrofit, conservation improvements are the best first steps you can take toward minimizing your energy use simply because they are more cost-effective. Solar heating is also a cost-effective conservation improvement, but on a ladder of economic priorities it follows the more basic improvements you can make to minimize your consumption of conventional fuels (gas or oil) or electric power. "Consumption" is also referred to as *load*, and this chapter is all about lightening that load. The amount of energy you get from the sun is a fixed amount; it doesn't vary with the number of gallons of hot water you use. All that changes is the amount of auxiliary energy you use, so cutting down on hot water means reducing gas, oil, coal or electrical hot water heating.

Minimizing your water heating load is easy to do. It involves simple once-and-done tasks that almost anyone can do successfully. There are basically two goals that you can work toward. You can reduce heat losses from various points in your hot water system, and, without sacrificing convenience, you can use less hot water. Working with these goals in mind you can easily cut your present water heating costs by 25 percent, and a 50 percent reduction is entirely possible. And with every improvement you make, energy is saved right away.

You'll also save money when you're ready to buy or build a solar water heater, because with a more efficient hot water system that needs less energy, you can meet a larger percentage of your remaining load with a smaller collector system. Let's say, for instance, that in a typical household of four people the annual water heating load adds up to 24 million Btu, or 24 MBtu. (A Btu, or British thermal unit, is a standard unit of measurement that, for one thing, helps with the comparison of different energy sources. For example, 24 MBtu is the equivalent of about 240 therms of natural gas or 7000 kilowatt-hours of electricity—a lot of energy, no matter where it comes from. One Btu equals the amount of energy needed to raise 1 pound of water 1 degree Fahrenheit. It also roughly equals the heat given off by a wood kitchen match that is burned completely.) You're hooked on the idea of going with a solar system, and you find out from this book or from a local dealer that in your region you'll need about 100 square feet of collector area to meet about 75 percent of that load. But wait! With conservation improvements you can knock that 24 MBtu load down to, say, 15 MBtu. You can meet 75 percent of that reduced load with just 60 square feet of collector area. The numbers speak clearly: At a very real price of $20 per square foot for the solar system (if you hire somebody to do it for you) your preconservation cost would be $2000; postconservation, you'd pay out just $1200.

Plugging Heat Leaks

The hot water systems in most North American homes were installed when energy was cheap and plentiful, and efficiency was, if anything, a minor concern. Yet with today's energy costs it's clear that too much hot water is allowed to pass through faucets and shower heads. Poorly insulated water heaters and uninsulated water pipes dissipate much of the stored heat. High thermostat settings mean that hot water is stored at unnecessarily high temperatures, resulting in high rates of heat loss. Fortunately, all of these energy wasters can be fixed.

Photos 1–1, 1–2: Thermostats for electric water heaters, above, and gas water heaters, right.

The heart of any hot water system is the electric or gas water heater. Some water heaters are oil fired, but they are used much less these days. Most electric water heaters have two heating elements, one near the top and one at the bottom of the tank. Each heating element is controlled by an adjustable thermostat, marked in degrees Fahrenheit, that is located under a removable cover plate. In a gas- or oil-fired heater the burner is at the bottom, and a flue pipe extends from the top. The temperature of a gas-fired heater is controlled by a thermostat at the lower end of the tank, near the burner. It typically has settings that read "warm" or "low" and "cool" or "vacation."

If your gas water heater's thermostat is set on "hot" or "high," the water may be as hot as 140° to 160°F. Often the factory setting on an electric heater is 140°F. That's too hot. Most people shower, for instance, at water temperatures below 115°F. You can save a lot of energy simply by turning the thermostat down. If you own an electric heater, you must

Photo 1–3: When insulating a gas water heater it's important not to get the insulation too close to the flue. Leave a gap of several inches.

first switch off the power (there should be a separate water heater circuit breaker in the main circuit box). Then remove the cover plates on the side of the heater. This will expose the adjustable thermostats. The upper thermostat should be set five degrees higher than the one at the bottom. For example, set the bottom thermostat at 110°F and the top at 115°F.

Once you have readjusted the heater's thermostat, reactivate its electricity, then leave it like that for several days. Gradually raise the temperature, up to 120°F, if you find that you need more hot water, or lower it if you want to find the minimum satisfactory settings. Naturally, the lower you set the thermostat the more energy you will save. With gas- and oil-fired heaters you only have to play with the thermostat lever. There's no need to shut off the fuel. The energy savings from thermostat setback can be substantial. For example, with a typical 52-gallon water heater a setback from 140° to 120°F annually saves 410 kilowatt-hours of electricity or about 17 therms of gas. At 7¢ per kilowatt-hour $28.70 is saved, and at 60¢ a therm $10.20 is saved—every year and all for a couple of minutes of no-cost conservation work. To carry it a little further, that much energy savings equals the annual output of up to 10 square feet of solar collector. Need we say more?

The next best improvement you can make is to wrap the heater in a blanket of fiberglass insulation. Most heaters have only 1 or 2 inches of fiberglass between the tank and the sheet metal cover. Adding 2 to 6 inches more is highly cost-effective at today's energy prices. You can either buy a water heater insulating kit (2 inches of fiberglass) from a hardware, plumbing or building supply store, or you can wrap the heater with 6-inch fiberglass insulation, which actually works out to be less expensive than the store-bought jacket.

Tie the insulation around the heater's sides with wire or cord and seal the seams with duct tape. Cut out a plug of insulation over each thermostat (on an electric heater) so you can have easy access. Be sure not to obstruct the pressure-and-temperature relief valve at the top of the heater. If you have a gas-fired heater, you must be sure not to cover up the combustion air intake vents at the base of the unit. You should also leave a gap of a couple of inches between the insulation and the flue pipe at the top of a gas or oil heater. It will cost $12 to $20 to buy the insulation necessary to properly wrap your water heater, but this 30- to 60-minute improvement can pay for itself in less than a year.

After the heater has been bundled in insulation you should insulate your hot water pipes. Whenever your household uses large quantities of hot water, lots of heat escapes through the pipes. To minimize this loss you should use pipe insulation that is rated to R-4 or better. (See the product listing in Appendix 1.) Cover only those hot water pipes that are exposed; it's not worth your time or money to rip open walls to insulate hidden piping. You should also insulate the cold supply pipe for 6 to 8 feet from the heater.

To save more energy some homeowners install timers on their water heaters. In many households most of the hot water gets used in one or two "peak periods" every day. If that's the case, there's really no need to keep the water heated to the maximum temperature 24 hours a day when it's really needed for only an hour or two. If your water draws are confined to two or three distinct periods a day, you can conserve energy by installing a timer. You'll still have hot water during the "off" periods because a well-insulated heater won't cool down that much when the elements are switched off. The timer is essentially a clock-driven switch that is rated for

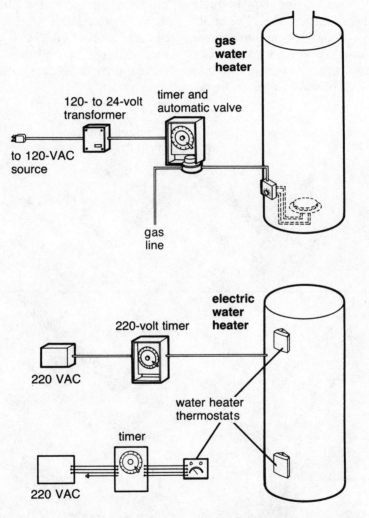

gas
water
heater

120- to 24-volt
transformer

timer and
automatic valve

to 120-VAC
source

gas
line

electric
water
heater

220-volt timer

220 VAC

water heater
thermostats

timer

220 VAC

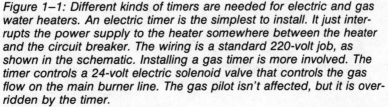

Figure 1–1: Different kinds of timers are needed for electric and gas
water heaters. An electric timer is the simplest to install. It just inter-
rupts the power supply to the heater somewhere between the heater
and the circuit breaker. The wiring is a standard 220-volt job, as
shown in the schematic. Installing a gas timer is more involved. The
timer controls a 24-volt electric solenoid valve that controls the gas
flow on the main burner line. The gas pilot isn't affected, but it is over-
ridden by the timer.

the same voltage as your electric-powered heater (usually 220 volts, sometimes 120 volts for small heaters). A timer is also available for gas-fired heaters. Figure 1–1 shows how these units are installed.

A gas-fired heater loses considerable amounts of heat through the flue. For about $50 you can buy an AmeriTherm flue damper that reduces this heat loss whenever the burner is off. The damper comes in various sizes to match the diameter of the flue pipe. Instal-

lation is usually pretty easy; just lift up the flue a few inches, slide in the damper and you're done. You may have to shorten part of the existing flue pipe, and you should be sure to install the damper in the direction indicated by the arrow on its label. There's no wiring required because this damper has bimetallic leaves that automatically open on temperature rise (burner on) and close on temperature drop (burner off).

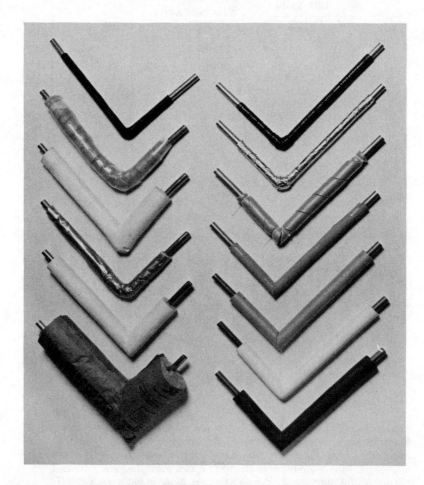

Photo 1–4: Many different kinds of pipe insulation are available to suit your exact needs.

Photo 1–5: A flue damper reduces the amount of heat that disappears up the stack of a gas water heater.

Using Fewer Gallons

Thermostat setback, tank and pipe insulation, timer and damper controls all work together to reduce heat losses, and they'll deliver significant energy savings. The next improvements are aimed at reducing the amount of hot water your household uses, and these, too, will return substantial savings. Reducing gallonage doesn't mean you'll have to cut back on the number of showers you take. In fact, these are changes that will mostly be unnoticed, except for the big savings on what you spend to heat water.

For starters, the utility that supplies your water may be piping it to your home at a pressure that is greater than necessary. Too much water is forced through your faucets and shower heads, and the water heater works overtime. You can find out what your water line pressure is from the water company. If

you have your own well there should be a pressure gauge near the pump or the pressure tank. If the pressure is above 45 pounds per square inch (psi) your water system could use a pressure-reducing valve. By lowering the pressure, which reduces the flow rate, you can save both hot and cold water.

A pressure-reducing valve can be bought at a plumbing supply store for $20 to $35. You'll also need a pressure gauge (cost: $5 to $10), which is installed downstream from the pressure reducer.

The gauge and the pressure-reducing valve are added to the plumbing near where the water main enters your house, but downstream of the main shutoff valve and the water meter (figure 1–2). There are different installation procedures for different types of piping. If your home has copper plumbing, you must sweat solder the valve in place, and for that you'll need the right soldering tools.

It's a good idea to learn how to solder copper tubing because once you do you can say good-bye to expensive visits from the plumber (see figure 1–3 for step-by-step instructions for sweat soldering). If you have steel pipes, you'll have to cut away a section long enough to accommodate a threaded reducing valve.

When all's well you can lower the water pressure. There is an adjusting nut on the valve that is turned with a wrench or screwdriver. Screwing the nut into the valve lowers the pressure, and while you're doing that, keep an eye on the pressure gauge. (Open a cold water tap to get the most accurate read-

ing.) A friend of mine has experimented with his water pressure and reports that it can be lowered to 30 psi or less with no loss of convenience. "Try 25 psi!" he encourages. What's the right pressure for your home? You can go as low as you like until you feel the flow is inconveniently slow. Of course, the lower the pressure, the lower your monthly water bill and the less water your heater will have to heat. Also, in some municipalities the sewer bill is figured on gallons used, so you can reduce that cost as well. There are other benefits: Lower pressure extends the life of your water heater, and "water hammer" problems

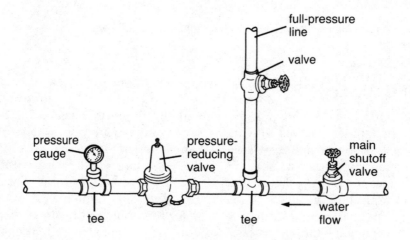

Figure 1–2: A pressure reducer is installed downstream of the main shutoff valve and the water meter. Keep in mind while looking for a place to install the reducer that it's a good idea to maintain a full-pressure line between the main shutoff valve and the reducer. When you're ready to install, close the main shutoff valve and cut into the water line. (It may help to drain the plumbing by opening the lowest spigot in your house.) Be careful when cutting the pipe, as you don't want to cut away too long a section. When working with copper pipe, you'll probably need to solder on two male adapters, which then get wrapped with Teflon joint tape to ensure a good seal. Then you can screw the valve onto both adapters. The pressure gauge should be installed a few inches downstream from the valve, using a standard tee fitting.

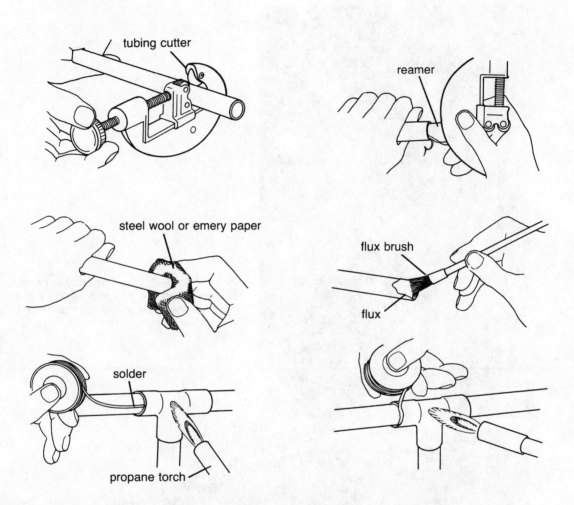

Figure 1–3: The right tools and a little practice make sweat soldering easy. Cut copper tubing with a tubing cutter (a hacksaw will also work but it makes a more jagged cut). Twirl the cutter around the pipe until the cutting disc digs in, then tighten the clamp again, until the disc cuts clean through the pipe. The cut end should have no jagged edges, inside or out. Scrape off any inside burrs with the reamer that is attached to most tubing cutters. Then clean the end with emery paper and/or steel wool to remove oxidized copper. Also clean the inside of the fitting (elbow, tee, or whatever) with a small round wire brush that's made just for this purpose. Next brush a thin coat of flux onto the polished joint and the fitting. Join the fitting and the pipe ends and heat the fitting with a propane torch. Don't bring the flame closer than an inch from the joint, and keep moving the flame back and forth, keeping it directed at the fitting. The joint is hot enough when solder held against the fitting melts. Keep laying on the solder and watch the melted solder flow in between the pipe and the fitting. When you see a shimmering ring of molten solder all around the joint, enough solder has been applied. Let the joint cool (the solder will lose its shininess) before moving it.

Photos 1–6, 1–7: Low-flow shower heads and faucet aerators are inexpensive, easy to install and save many gallons of hot water every day.

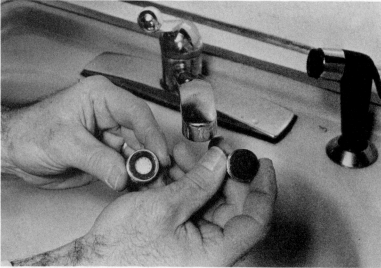

(those annoying rumbles that shake plumbing), if you have them, are reduced. A final note on pressure reduction: It's a good idea to leave some outside garden spigots at full pressure. God forbid, but if there's ever a fire, you'll want to be able to shoot water far and high. You can create a full-pressure line by teeing into your water main upstream of the pressure reducer (figure 1–2).

Probably the most cost-effective way to minimize gallonage is by retrofitting flow-restricting shower heads and faucet aerators. The shower head that's in your bathroom now probably passes water at a rate of 3 to 6 gallons a minute. (You can find your flow rate by running the shower at its normal rate for a minute, filling a container and then measuring how much was collected.) A flow-restricting shower head, which sells for $10 to $20, cuts water consumption down to about 2 gallons a minute, depending on the pressure. This means that a five-minute shower that used to consume 20 to 30 gallons of water will now use about 10 gallons.

I've been using a flow-restricting shower head for two years. I'm the kind of person who likes to hang around in the shower thinking and singing, so I consider myself pretty critical when it comes to shower quality. The first time I used it I expected that sharp needles of water would blast me across the tub, but I discovered instead that the shower head delivered a powerful but comfortable spray. Quickly my hair was wet and hot water was streaming down my neck and shoulders. The next thing I knew I was reaching for the soap and singing in the shower! Now whenever I visit friends who have old-fashioned shower heads I'm amazed at the amount of hot water that's needlessly wasted.

Flow-restricting shower heads make lots of sense and they're easy to install. Simply unscrew the old unit, wrap some Teflon pipe tape around the exposed threads, screw on the new shower head and, presto, you're saving plenty with every shower. Low-flow aerators are screwed in place by hand and should be installed in every household faucet, from the kitchen to the bathroom.

After showers and tub baths one of the biggest hot water users in households is the washing machine. Your washer probably has a cold water wash cycle, and energy can be saved simply by throwing a switch. Cold water detergents clean clothes well and save 12 to 25 hot gallons with every load.

Going Tankless

So far we've concentrated on improving the efficiency of a hot water system by reducing heat losses and by using less hot water. Another anti-heat loss strategy involves replacing your tank-type heater with what is called a tankless, or demand-type, water heater, or using a tankless unit with your present water heater.

Even after you've turned down a tank heater's thermostat and wrapped it with insulation, some heat is still lost from storage and in the hot water line (distribution losses). In some cases, it can make more sense to heat water only when it's needed, and that's what tankless heaters do. They don't heat a drop of water until a hot water faucet is opened and the water flow turns on a switch inside. Water passes through the unit once and then flows straight to the point of use. In this way no expensive heat is left to cool its heels in a storage tank, waiting for the next hot water draw.

Depending on your hot water needs, one or two or three tankless water heaters can completely replace a tank heater. Typically they are installed near the places where hot water is needed, such as in the kitchen, bathroom and laundry. In the kitchen a tankless heater can work with a tank heater to save energy. If you have a dishwasher that needs 140°F water, you can keep the tank water heater setting at 115°F and install a tankless unit on the water line that supplies the dishwasher. (Some diswashers have their own built-in heaters, which would obviate the need for a tankless unit.)

Tankless heaters also help to reduce distribution losses when there are long (over 50 feet) runs between the tank heater and other points of use. Instead of piping hot water across that distance you can install a tankless heater right where the water is needed. With new construction a tankless heating system also eliminates the need for separate hot water lines since all that's needed is a cold supply to the heater.

There are many kinds of tankless water heaters on the market, from midget kitchen sink models that instantly produce a cup of hot water to very powerful electric or gas units that can supply a household's entire hot water needs. They are rated according to the amount of energy they use, either in kilowatts (electric) or in Btu's per hour (gas). The outlet temperature of the water provided by a tankless heater is determined by both the flow rate and the temperature of the inlet water.

Photos 1–8, 1–9: Gas or electric tankless water heaters eliminate the need for long hot water pipe runs. When one is installed next to the dishwasher, the temperature of the rest of the household's hot water can be reduced because the tank water heater thermostat can be set below 140°F.

Thus, along with their power ratings, tankless heaters can be compared by the temperature rise they produce at different flow rates (these specifications should appear in the manufacturer's literature). As you might imagine, the slower the flow rate, the higher the outlet temperature, which is why it is best to use them with flow-restricting shower heads and faucet aerators. A look at table 1–1 shows that the temperature rise through various electric and gas-fired tankless heaters changes greatly with changes in the flow rate and power input (kilowatts). Even if the measured flow rates through various points of use are within the recommended 1 to 2 gallons per minute (gpm) range it's a good idea to retain the tank water heater as a low-temperature preheater for the tankless units. The tank heater could be set as low as 80°F to guarantee an adequate temperature rise through the tankless unit. Another reason for keeping the tank heater

TABLE 1–1

Power Ratings (in Kilowatts) for Tankless Water Heaters Based on Required Flow Rate and Temperature Rise

Tap Flow Rate (gpm)	Temperature Rise (°F)				
	20	40	60	80	100
½	1.6	3.2	4.8	6.4	8.0
1	3.2	6.4	9.6	12.8	16.0
1½	4.8	9.6	14.4	19.2	●
2	6.4	12.8	19.4	●	●
2½	8.0	16.0	●	●	●
3	9.6	19.2	●	●	●

● In order to have both a high flow rate and a high temperature rise, two or more tankless units would have to be installed in series.

is that it could be used to store solar-heated water. Even a 40-gallon tank can be used with 20 to 40 square feet of solar collector, depending on the abundance of your solar resource.

Should you decide in favor of a tankless heater, don't get rid of your present water heater if you're also thinking about heating water with a solar collector or a woodstove. Because the sun doesn't shine all the time and since you may not use your woodstove 24 hours a day, you'll need a hot water storage tank. Your present water heater can probably be used for this, if it's at least a 40- to 50-gallon unit.

Some homes would actually be better off without an old-fashioned version of tankless water heating. In many houses domestic water is heated by the same oil- or gas-fired boiler that heats the home in winter. In a system like this a pipe coil heat exchanger passes through the boiler water containment. The coil is connected to the house's cold water supply, and it is heated as it makes a single pass through the 180° to 200°F boiler water. Commonly called the "summer-winter hookup," this might be an efficient way to heat water in the winter, when the home must be heated anyway, but come summer it makes no sense to operate a boiler just for domestic water heating. A boiler providing both space heat and domestic hot water can operate at 70 to 80 percent efficiency. But when a boiler is used for domestic water heating only, the efficiency can drop to as low as 15 percent, which translates into a substantial increase in fuel costs. Under these circumstances installing a separate gas or electric tank heater would be a cost-effective investment. In winter the water would still be heated by the boiler, but as soon as the need for space heat ended, the boiler would be shut off and the tank heater would work along at higher efficiency (figure 1–4). The fuel savings from nonuse of the boiler will pay for the new tank heater in two to four years.

Another option uses a plain pressure tank (no water-heating capability) as a *tempering tank* to raise the temperature of the water before it enters the water heater. The incoming water is usually in the 40° to 55°F range as it enters the house, and by allowing it to sit in the uninsulated tempering tank it warms up to the temperature of the room. Thus, a tempering tank makes sense only if it's installed in a spot that is usually warmer than the incoming water. A 70°F room means a

15- to 30-degree temperature rise. That's a good bit of energy saving by giving the water heater warmer water. The benefit comes primarily during warm weather when the room isn't being heated by the house space heating system.

Greywater Anyone?

One rather different conservation technique that could save a lot of energy is *greywater heat recovery*. Greywater is what gurgles down the drain after you've showered or washed your clothes (*blackwater* is what gets flushed down the toilet). Warm or hot greywater is dirty and can't be used again,

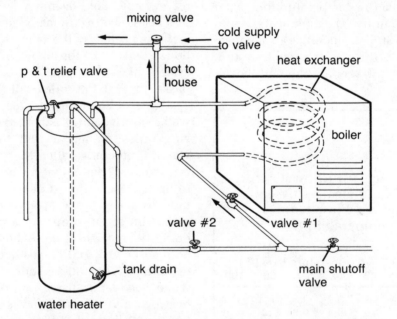

Figure 1–4: In some homes the space-heating boiler contains a heat exchanger for heating domestic water. This is a good feature when it's used during the space-heating season, but it's very inefficient to use it in warm weather. You'll actually save money in two or three years if you install an electric water heater for use when there is no space heating. Valve #1 is opened in the winter and valve #2 closed; when the space-heating season ends, the valve positions are reversed.

but why should all that expensive heat it contains go down the drain? With a heat exchange system some of that heat could be recycled and used again. There has been only a small amount of research done to date on this idea, but greywater heat recovery is basically very simple.

Heated greywater is diverted into an insulated holding tank from the bathtub, shower, kitchen sink, dishwasher and clothes washer (or some combination of these inputs) before it enters the sewer line. This tank contains a pressurized pipe coil heat exchanger that's located on the cold supply line to the existing water heater. When a hot water outlet is opened, cold water flows through the heat exchanger coil, where it is preheated before entering the water heater. Of course, the water in the coil never touches the greywater; it only absorbs heat that otherwise would be lost. Studies have shown that a greywater system can recover from 30 to as much as 50 percent of the total energy used for heating water. But, as mentioned, there has been little development work done on this idea and you can't buy a ready-made system. It is possible at this time, though, to make one with readily available materials. Figure 1–5 shows an experimental design that some do-it-yourselfers have had success with, although we haven't built and tested this or any other designs ourselves. It should be noted that designing a retrofit greywater heat recovery system is going to present difficulties that may be too formidable for the average home craftsman, especially with the sanitation aspects of such a system. Another basic difficulty is likely to be separating greywater (sinks, showers, tubs, dishwasher) from blackwater (toilets, bidets) sources, since they usually flow through the same soil pipes. In short, while the idea has merit, little work has been

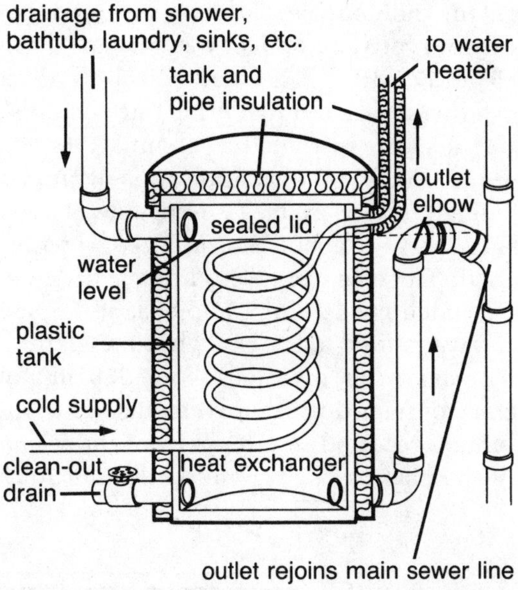

Figure 1–5: A greywater heat recovery system can recapture valuable Btu's that would otherwise disappear down the drain. Here is an idea for a heat recovery tank that uses a 30- to 55-gallon container that is fitted with a heat exchanger and a greywater inlet and outlet. Water from the shower and tub, the sink or dishwasher and the laundry is diverted to the tank. Cold water, on its way to the water heater, is preheated when it passes through the heat exchanger. The tank lid should be sealed, but removable. The tank and the heat exchanger outlet should be insulated.

done to develop a safe, practical system. (An article in the August 1981 issue of *Solar Age* magazine discusses a variety of design possibilities.)

Conservation: Cheaper Than Heat

These days an average family of four spends from $200 to $450 every year to heat water. As much as half that cost can be eliminated with the methods that have been discussed. These conservation improvements can pay for themselves in just one or two years.

One homeowner I know reduced his yearly Btu consumption for water heating from 25.5 to 17 MBtu. The water heater's thermostat was turned down from 130° to 120°F. Pipe runs were insulated for about $7.50. Ten dollars' worth of insulation was wrapped around the water heater. Two low-flow shower heads, costing about $10 apiece, were installed. Three faucets were fitted with aerators, which cost about $2 apiece, and a $35 timer was added to the water heater. All told, the homeowner invested about $78 in hot water conservation improvements. This investment resulted in a 33 percent reduction in water heating costs. At 5¢ a kilowatt-hour,

the annual water heating bill was reduced from about $375 to $250. The $78 investment was repaid in less than a year.

Some conservation improvements are more cost-effective than others. It costs nothing, for example, to turn down a water heater's thermostat, and savings are realized immediately. On the other hand, a tankless water heater will cost several hundred dollars and may not pay for itself for several years. Conservation improvements should be viewed as an investment, and, as with any other investment, consider first those measures that cost the least but repay the most in the shortest amount of time.

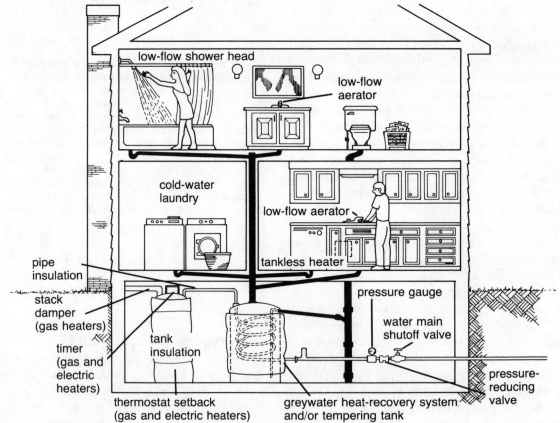

Figure 1–6: Most hot water conservation improvements are inexpensive and pay for themselves quickly. A water-conserving house is also less expensive to solarize. This illustration depicts improvements that can be made in most homes.

2

SOLAR WATER HEATING SYSTEMS: AN INTRODUCTION

Just as there are different house styles and different climates, there are also different kinds of solar water heating systems. This is fortunate because no single type is best for every house or climate. In fact, there are usually two, three or more systems that would work in a given situation, and homeowners are faced with a decision to select one that is the best.

Choosing the best system for your home depends on several variables, such as the climate and the possible need for freeze protection, the cost of the system, whether you want to build all or part of the system yourself, the physical layout of your house and yard, the structural integrity of your roof, aesthetics, the ease of operation and the maintenance requirements of different systems, and

Photo 2–1: Here is a turn-of-the-century batch heater. The tank was painted black and slanted to face the sun. By midafternoon it could heat a batch of water, which had to be used quickly or the heat would be lost to the night.

your hot water consumption. This chapter and the next explain how you can pick the right system for your home by keeping these "solar variables" in mind, but for starters you'll want to become familiar with some of the basic concepts and language of solar energy, and also with some of the history of how solar water heating has evolved over decades of use in the United States and other countries.

It has often been said that there's nothing new under the sun, but over the years count-

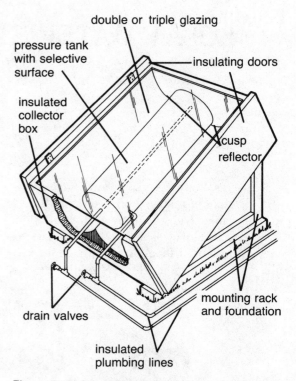

Figure 2–2: A batch collector is a model of simplicity: A black tank is put inside a glazed, insulated box and left to roast in the sun.

Figure 2–1: An old advertisement from Arizona *magazine, circa 1913, extols the many virtues of a solar water heater.*

less devices have been invented to heat water with sunlight. Some ideas have worked, and there has certainly never been a shortage of ideas. In this country, the first solar water heaters—produced in the late nineteenth century—were no more than blackened metal tanks placed in the sunshine. These primitive collectors worked, but without sunshine they quickly lost heat. Over the years more efficient and reliable collectors evolved, with a major improvement being the addition of glazing over the south-facing side of a collector box.

Two Collectors for Many Systems

Today, two basic kinds of collectors are used widely across North America: *batch* and *flat plate*. As you'll see in later chapters, there are many variations on these two types, and there are also several kinds of systems that use a batch or flat plate collector component.

A batch collector is a large, insulated box that encloses one or more 30- to 50-gallon tanks (figure 2–2). The tanks are painted black to absorb a maximum amount of solar energy. Often the inside of the box is covered with a reflective material such as aluminum foil or aluminized Mylar, to bounce more sunlight onto the tank. A double or triple layer of glass or plastic glazing covers one side of the box to admit sunlight and hold in heat. Sunlight passes through the glazing and strikes the tanks, heating water that's stored there. Solar-heated water is drawn from the batch collector for use whenever a hot water tap is

opened. This is because the tanks are plumbed into the cold water line supplying the existing water heater. The batch tank feeds the water heater, which then provides hot water on demand. Depending on the climate and the season, a batch heater serves as either a preheater (in the winter) to the water heater or as a total supplier (in the summer) of hot water.

If you've ever left a garden hose on a sunny lawn for an afternoon and then came back to find the hose filled with hot water, you already have a pretty good idea of how flat plate collectors work. A flat plate collector (figure 2–3) is a shallow box containing a sheet of blackened metal, usually copper, to which tubing, also copper, is bonded. Sunlight striking the absorber becomes heat, and as cooler water is circulated through the tubing the heat migrates to the water. The heated water then flows into an insulated storage tank inside the house, where it waits to be used.

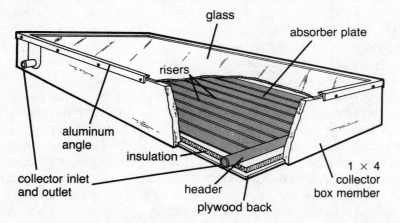

Figure 2–3: Flat plate collectors are pretty simple, too: an insulated box, glass, and a copper absorber plate.

Some Solar Words and Ideas

Before we go any further, there's a need to introduce some of the basics of solar language, which will lead to a clearer understanding of just how solar energy works. As we'll see, there's a good bit more to it than just having a sunny day.

Like most engineering terminology, the language of solar energy is very literal. Energy radiating from the sun is called *solar radiation*. Incoming solar radiation that falls on earth's surface is squeezed into the word *insolation*. *Absorption* is what happens when solar energy is absorbed, as heat, by a dark surface, such as an absorber plate or a car seat. *Transmission* occurs when sunlight passes through a transparent material, such as collector glazing. The presence of glazing is critical to the performance of any type of collector. It not only transmits the energy but, more important, when the energy becomes heat, the glazing traps it inside the collector. This is commonly called the *greenhouse effect*, quite noticeable if you've ever been in a greenhouse or suffered the summer's heat inside a closed-up car. Without glazing, solar domestic water heating just wouldn't be worth it. In later chapters you'll see that there are many kinds of glazing and that they can be used in a number of configurations to optimize collector performance.

Once solar energy has been collected, the resulting heat travels by *conduction* to the cooler water inside a batch tank or in the tubes of a flat plate unit. Conduction through a solid material is one of three ways that heat is transferred. Heat also travels through space by *radiation*. A hot surface, like that of a woodstove, is said to radiate infrared heat to other, cooler surfaces. Infrared radiation, like solar radiation, doesn't become sensible heat

until it actually strikes another surface. When you face a hot woodstove or open fire, your skin is heated by radiation. Pass your hand in front of your face, and you block that radiation. Your hand heats up; your face cools down. The third form of heat transfer is by *convection*, in which heat is moved by a fluid such as air or water. As the old saying goes, heat rises, and it does so by *natural convection*. Heated air is lighter and more buoyant than cooler air, so it rises. Heated air can also be blown around by a fan, like when you want to get it out of your attic in summer. This is called *forced convection*. The water in a flat plate collector is also a convective medium. It can be circulated to and from a storage tank (figure 2–4) simply by the heating action of the sun. Hot water rises, while cooler, denser water falls, all by natural convection or *thermosiphon* flow. When a pump is used to push water around, the collector system operates by forced convection.

For higher efficiency, water (or antifreeze) is most often circulated through a flat plate collector by a thermostatically controlled pump, and the system is said to be *active*. When no pump is used (thermosiphon flow) the system is *passive*. A batch collector system, in which pumps are never used, is another form of a passive system.

Just above it was mentioned that antifreeze is used in some systems as a heat transfer medium that also provides freeze protection. You can't of course drink the stuff (it's too expensive, and it makes you ill), but you can send it through a *heat exchanger* to transfer solar heat (by conduction) to potable water in a storage tank. This is done in an *unpressurized* (no pressure other than atmospheric pressure), *closed-loop* system (figure 2–4). In a *pressurized* (atmospheric plus house water line pressure), *open-loop* system (figure 2–4)

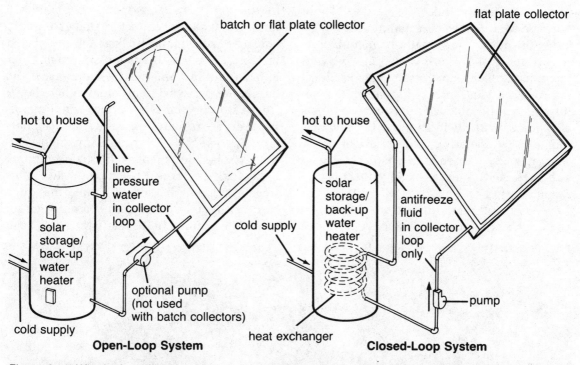

Figure 2–4: What's the difference between a closed-loop and an open-loop water heating system? In an open-loop system domestic water is heated directly by the solar collectors; batch, thermosiphon and draindown are examples of open-loop systems. A closed-loop system uses a heat exchanger to pass heat from the collector loop to the domestic water, as in drainback and antifreeze systems.

water (never antifreeze) is supplied by a direct link with your house's pressurized water system. Potable water thus circulates through the collector to the storage tank, with no need for a heat exchanger.

Using antifreeze in a closed-loop system is but one of several ways of having *freeze protection* for collectors and exposed plumbing. Batch systems and water-filled flat plate systems both must be safeguarded from below-freezing temperatures, and as you'll see in a later chapter, the differences among various system designs are mostly due to different solutions to the need for freeze protection.

This batch of words will suffice for a while; there'll be a few more coming at you later. Before delving into the aforementioned list of variables, a little discussion of what's going on in the solar industry will give you a look at where solar technology is heading, and why.

From Small Beginnings, a Huge Solar Industry

The solar collector may have been narrowed down to two generic types, but the industry that is comprised of hundreds of solar manufacturers is complex and ever changing. America's solar industry today is competitive, risky and quite diverse. There is a regular stream of new manufacturers entering the market and would-be manufacturers dropping out. There is also much dynamism in the technology itself. New products improve old ideas or provide innovative solutions to old problems. Some new products also bomb. With its technology the industry seems to be operating in a couple of directions at once—toward both greater simplicity and increased complexity in collector and system design.

Many solar equipment manufacturers advocate simple, passive systems. Taking this to the extreme, some manufacturers have marketed Japanese collectors that essentially consist of rugged plastic bags set into glazed boxes. Critics say this is too simple and not appropriate for American consumers. If the bags aren't protected from cold weather, they'll freeze up. On the other hand, some manufacturers are responding to the American passion for gadgetry by producing high-technology solar equipment such as concentrating collectors. In these collectors, water, antifreeze or gas is passed in a tube that runs through the middle of a transparent vacuum tube. The vacuum helps retain the sun's heat.

Photo 2–2: Concentrating collectors can heat water up to 500° or 600°F. Used mostly by businesses that need plenty of inexpensive, relatively high-temperature water, they usually aren't the best choice for homeowners.

Photos 2–3, 2–4: Solar water heaters are popular all around the world. Made from materials at hand, they often reflect the technological prowess of the countries where they were made. Left, the collectors atop a hotel in Xian, China, are supported by rough-hewn beams, while a Japanese collector, below, shows more modern design techniques.

Each tube is surrounded by reflectors that concentrate more solar energy on the tubes. Often the collectors are motorized to track the path of the sun, like a sunflower does. A sunflower, however, has no moving parts, but a collector like this has many, all of which must work together. Critics say this is excessively complex, especially for meeting the water heating needs of a typical home. Such collectors also produce water temperatures of 180°F and higher, so they're actually too powerful for the needs of most homes.

American solar entrepreneurs aren't the only ones who are following multiple paths in the development of new technology. The Chinese, who do not have much money for fancy hardware, are developing simple passive systems that can be built with wood and other readily available materials. The Japanese, who have more capital and much more readily available technology, are pursuing a higher technology route. Some collectors made in Japan are truly state-of-the-art equipment, with trade names like Panasonic and Sanyo.

Thousands of miles away, in Israel, very simple, passive solar water heaters have been popular since the early 1950s. Passive systems rarely have automatic freeze protection, and since that's not a concern in the Israeli climate, there's no problem. But in most of North America there is, of course, an absolute need for freeze protection. Consequently, some types of passive systems must be completely shut down and drained in winter to protect against freezing damage, although there are ways that they can be allowed to operate safely in the coldest temperatures.

You're probably starting to see the point. Across the globe, though many different kinds of systems are available, people tend to pick the one that best fits their needs, their climate conditions and their pocketbooks. This is true

in America, too. The industry sells a wide variety of systems for home use, from simple plastic bags to complex vacuum-tube schemes. Between these two extremes are flat plate and batch collectors, which are the most popular kinds of solar water heaters used in America because they best fit the needs and finances of the majority of households. But even with the narrowing of choices to two collector types, there are still over half a dozen systems that can be used with them, plus little variations on system designs that give you even more options for customizing the one that's right for you. It's especially important for you to know what your home needs before you go shopping.

Sorting Out the Variables

As was stated earlier, there isn't one system that's right for every house everywhere. There are just too many variables for one all-encompassing system design. But you can work with the variables to make a prudent choice from among the systems presented.

To Buy or To Build?

Perhaps the most important variable to consider is the amount of money you want to spend. A turn-of-the-century solar water heater advertisement proclaimed that "sunshine, like salvation, is free," but the truth is that a good solar water heating system can be very expensive. The price for having an experienced solar contractor do a turn-key installation of one of the nationally known flat plate systems could easily be $3000 to $4000, before any tax credits are taken (tax credits are discussed in the next chapter). The system would probably live up to the manufacturer's performance claims and have a long, trouble-free life. A commercially made batch collector can cost $1200 or more. In either case, if you can afford that kind of money, and if you're not inclined toward building your own system, you can shop for a commercial system, which can be confusing. Why? There are presently several hundred manufacturers of various components of solar water heating systems. A recent solar industry catalog of parts listed 187 different models of flat plate collectors, made by almost as many manufacturers. Which to choose? The good news is that you won't be exposed to that much variety. Since you'll probably be looking for dealers and installers who are close to home, you won't have nearly so many brands of collectors to choose from. (Check your Yellow Pages under "Solar," and start dialing. There are also a number of mail-order outlets for solar components, some of which are listed in Appendix 1.)

Speaking very generally, most collectors operating in the same climate conditions can capture and deliver about the same amount of energy from the sun. There are performance differences, but the primary differences are in price and in the materials they've been built with. In your shopping you want to find a system that's right for your home, that's durable and reliable and that delivers the most energy for your money. In short, you're shopping for the best buy (which is not always the best price). With so many products on the market it becomes especially important that you know what you want before you go shopping.

As a system buyer it's likely that you'll also be in the market for an installer. Installers usually have a certain water heating system that they offer to all prospective customers, and because of that you can't assume that you'll get the most objective advice from a solar contractor. This person probably feels that his system is the best for you, bar none.

Again, that's where this book will give you the background you need in order to talk with as well as listen to a contractor. Chapter 9 gets into more of the ins and outs of working with a contractor. As a system buyer, as a consumer, you'll be much better off with knowledge.

Home-Built Systems and Savings

If you have reasonably good building skills, you can save lots of money by building a system yourself. A homemade batch heater can cost as little as $300 or $400, and you can save at least $1000 that would otherwise go to a contractor by building your own flat plate system. It's certainly a way to go, and in following chapters you'll read step-by-step instructions for design and construction. It's not an impossibly difficult task, but don't kid yourself—it's not child's play, either. You must have good expertise in carpentry, plumbing and electrical work. You can probably build a collector if you have ever built a stud wall. If you've ever wired an electrical box, you can most likely handle the necessary electrical connections. And if you've spent some time sweat soldering pipe and doing other plumbing chores, you can certainly succeed with your solar plumbing. But be sure you can do the job right; mistakes, which could lead to a frozen flat plate, can be expensive. Of course, you can always learn new skills. If you're hot to do something solar, well, go ahead. But take your time. Practice different tasks before committing your newly acquired skills to the actual installation.

As a less experienced do-it-yourselfer, you might just want to install a store-bought collector or heat exchanger tank to save time and effort. Or you can simply design and specify the system of your choice and then contract with someone else to install it. But

Photo 2–5: An absorber plate is laid into a collector box.

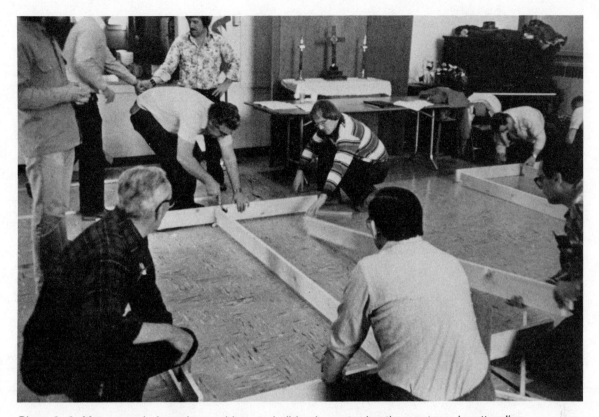

Photo 2–6: Many people have learned how to build solar water heating systems by attending workshops.

if you really want to gain some solar skills, a good way is to sign up for workshops and seminars. Many good do-it-yourself solar workshops are held throughout the United States and Canada, and if you keep a lookout you may find one in your area. There is probably a nonprofit solar energy association in your region that can fill you in on coming events. Some major associations are listed in Appendix 2.

Solar Access

Your building skills and your finances aren't the only things you must consider in your planning. Your house's physical layout and exposure to the sun naturally play an important role in determining the right solar water heater for your needs. It's obvious that your collectors will have to be placed where there's plenty of sunshine. Needless to say, a $3000, state-of-the-art flat plate system is worthless beneath the shade of the old oak tree. In general, the more sunshine you have access to, the more options you have for collector placement and system design.

In the next chapter you'll read how to conduct a site assessment of your home to determine how much solar energy is avail-

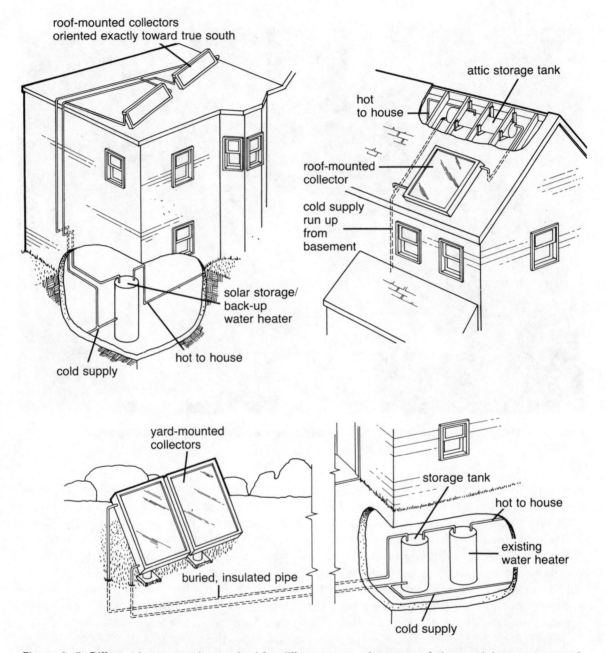

Figure 2–5: Different homes are best suited for different types of systems. A thermosiphon system such as the one shown at top right requires a storage tank at least 18 inches above the collectors. Pump-powered active systems make it possible to keep the storage tank in the basement and give the installer more options for collector placement. At top left are roof-mounted, active collectors. Active systems also make it possible to install the collectors on the lawn away from the house, as shown at bottom, to overcome possible shading problems.

able at your site. This assessment will heavily influence your decision about the type and size of collector that's right for your house. Another related consideration is whether or not the collectors can be oriented toward the south and if they can be tilted at the appropriate angle for optimum energy collection. (In later chapters you'll learn how to install collectors at the right tilt and compass angles.)

The layout of your house might also prevent the installation of a thermosiphon system. Because flat plate thermosiphon systems depend on the ability of hot water to rise, the bottom of the hot water storage tank must be at least 18 inches above the top of the collector, usually in the attic if the collectors are roof mounted. If you don't have an attic, or if your attic can't support the weight of a 500-pound water tank, you might have to install an active system that allows placement of the tank below the collectors. A filled water tank typically weighs 400 to 600 pounds. The key to a safe attic installation is to have the tank supported by a load-bearing wall or to use heavy planks (2 × 6's on edge) to spread the weight of the tank over several joists in the attic floor. You can reinforce attic joists and rafters by doubling them up with boards of the same dimensions.

The same consideration must be given

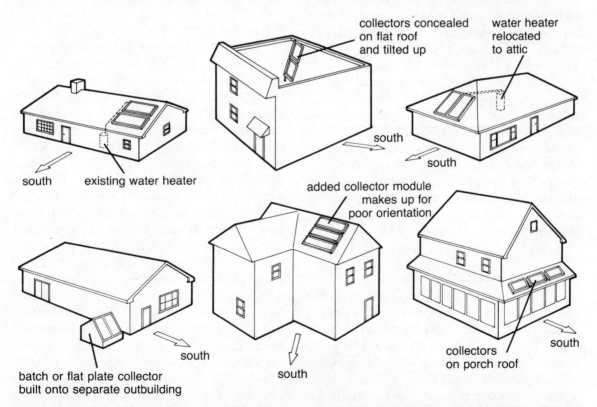

Figure 2–6: We've all seen flowers pointing at the sun, looking beautiful while collecting energy. That's the idea behind good collector placement. Collectors should face true south; they should be slanted to the proper tilt angle to collect as much energy as possible year-round, and they should blend well with the look of the house.

to a roof-mounted batch system. Typically, the roof rafters will need reinforcement, and, if that's not possible, the collector will have to be installed elsewhere, like on the lawn. You may have to switch to much lighter flat plate collectors, which are suitable for most roof structures (figure 2–5).

Good Looks

Aesthetics are clearly an important consideration. All collectors work best when facing true south or at least within 30 degrees east or west of south, and this requirement will certainly affect the positioning of the collectors. If you have more than one south-facing location available, appearances should be carefully considered. Some people try to conceal their collectors from street or yard view, which is fine and is often a good strategy for flat roofs. In other cases the collectors can only be mounted onto an exposed roof pitch, in which case their positioning relative to the roof dimensions, a chimney, a dormer or gable, should be considered for balance or symmetry. Color is important, too. With flat plates, for example, the collector box can be stained or painted to match roof color. For exposed collectors, the ultimate goal is enhancement of a house's total look, not disruption.

Après Installation

Another important consideration is the amount of time you want to spend attending to your solar water heater once it's installed. Some flat plate systems are completely automated and need little maintenance. Some batch heater designs, however, include insulated reflective doors or hatches that must be opened in the morning to catch the optimum amount of sunshine and then closed at night to hold in heat.

Make no mistake about it: Every solar water heating system, whether simple or complex, homemade or store-bought, passive or active, needs periodic maintenance. Every so often a system must be checked for leaks, or a collector must be repainted, or freeze protection devices must be tested. If your solar water heater was installed by a dealer or contractor, you may be able to purchase a maintenance contract, and those chores will be done for you. Of course, if you do it all yourself, you automatically assume all maintenance responsibilities. When the various systems are fully described later on, you'll find out more about the operational and maintenance requirements of each.

F-f-freeze Protection

Depending on the climate, you might have to perform special tasks every winter to ensure that the water inside your solar collector won't freeze. Freeze protection can be fully automatic in some systems, but sometimes an exposed collector (such as a batch or a thermosiphon flat plate) must be safeguarded from freezing by the more drastic measure of draining it and not using it in the wintertime, allowing it to hibernate until spring. You won't have to worry about freeze protection if you live in more tropical climes, but if you live in most areas of the United States, freeze protection is a primary consideration. It's also a major factor in system design. In fact, all the active systems described in this book are different primarily because of their different approaches to freeze protection. One approach is the *antifreeze system*, which is a closed-loop design that circulates antifreeze through the collector. A heat exchanger is used to transfer heat to stored potable water. In an open-loop *recirculation system* a *freeze sensor* is used to turn on the solar pump when air temperatures dip to 38° to 40°F. The pump

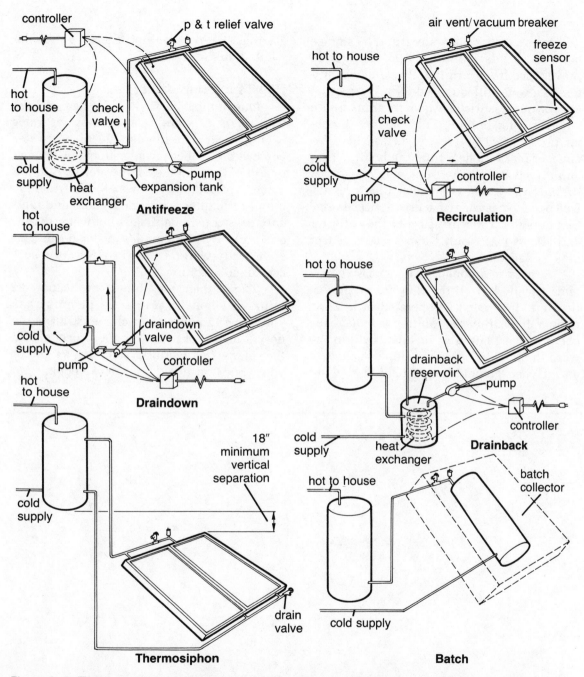

Figure 2–7: The six systems represented in this illustration are antifreeze, recirculation, draindown, drainback, thermosiphon and batch. Except for the recirculation system, which should only be used in Sun Belt climates, they all can be installed throughout North America. Many homes can accommodate more than one type of system. A particular type of system is often chosen because of other variables such as personal preference, cost or parts availability.

"borrows" a little hot water from the storage tank and sends it through the collector to prevent ice from forming. A *drainback system* is a water-filled closed-loop system that allows unpressurized water to drain into a well-insulated holding tank whenever the solar pump turns off. A pressurized heat exchanger picks up heat from the holding tank and directs it to the existing water heater. A *draindown system* is a water-filled pressurized open-loop system that uses a special valve that cuts the water pressure to the collector and allows it to drain when a freeze sensor detects danger.

As was mentioned earlier, passive systems usually have little or no freeze protection, and they can't be operated in freezing temperatures. But one solution is to have the batch or flat plate unit installed within the warmth of the building envelope, such as in an attached greenhouse or behind skylight glazing. That way nothing freezes, unless the whole house does!

How Much Hot Water?

One of the biggest variables that will determine the type and size of your solar water heating system is the amount of hot water you need. You'll be given instructions on how to add up your home's hot water load in the next chapter. If you followed the instructions found in chapter 1, your household undoubtedly uses considerably less hot water than it once did. By reducing hot water needs you are already well on your way toward a successful solarization.

These then are things you can start thinking about in your solar planning. As you'll see in subsequent chapters, it is now only a matter of matching those needs with the solar water heating system that's right for you, your house and your pocketbook.

3

CHOOSING THE RIGHT SYSTEM

By making the proper choices of system type and collector and storage sizing and placement, a solar water heating system can be tailor-fitted to a home. As was explained in the last chapter, there's no mystery to the decision-making process, just a number of variables that should be considered. This chapter will go deeper into these variables, and we'll actually go through a couple of decision-making efforts undertaken by two fictional families with unusual names: the Joneses and the Smiths. This chapter also includes the how-to of a simple calculation system you can use to determine how much energy can be produced at your site by the system you choose, and how much that energy is worth.

How Many Gallons?

All solar water heating systems have two basic components, the collectors and a hot water storage tank. (In batch systems the storage tank is also part of the collector.) You naturally want to size these components correctly, and to do that your first task is to estimate your hot water usage.

It would be simple to determine your home's hot water usage if a meter were attached to your water heater, but since hot water isn't metered you've got to look at all your various uses and add up some averages.

The estimation becomes easy when you categorize the different ways you use hot water. In most American homes hot water is used for dish washing, clothes washing,

household cleaning, food preparation, hand and face washing, shaving, showering and bathing. The typical amounts of hot water drawn for each purpose are listed in table 3–1.

When you add up your gallons, be methodical; make accurate counts on things like daily showers and baths, laundry loads and

Table 3–1	
Typical Hot Water Consumption	
Automatic washing machine, hot cycle	21 gal./load
Automatic washing machine, warm cycle	11 gal./load
Automatic washing machine, cold cycle	0 gal./load
Automatic dishwasher	15 gal./load
Dish washing by hand	4 gal./load
Food preparation (4 people)	3 gal./day
Household cleaning (4 people)	2 gal./day
Hand and face washing	2 gal./day/person
Wet shaving	2 gal./day/person
Tub bath	15 gal./bath
Showering, regular shower head	25 gal./shower
Showering, low-flow shower head	12 gal./shower

SOURCE: Joe Carter, ed., *Solarizing Your Present Home* (Emmaus, Pa.: Rodale Press, 1981).

so forth. As the table shows, you can figure that each person uses about 2 gallons of hot water daily for hand and face washing. If you have a dishwasher, determine the number of loads it washes each day, and multiply it by 15 gallons per load.

When figuring your laundry hot water use, a good method is to estimate the average number of loads washed each week instead of each day. For instance, if you wash four full loads a week in a top-loading washer that's set on the hot water cycle, about 100 gallons of hot water are used weekly. To find the daily hot water usage, simply divide the weekly average by seven, which in this case would come to about 14 gallons.

Continue adding up your various hot water uses until you've included them all. If you followed the conservation suggestions in chapter 1, each person in your house probably uses around 10 gallons of hot water a day. In a nonconserving household, however, daily consumption is closer to 20 gallons or more per person. The cost implications there are pretty obvious.

The next thing to check is the size of your existing water heater. Typical sizes for electric water heaters are 52, 65 and 80 gallons, all of which are candidates for double-duty service in storing solar-heated water. Typical sizes for gas-fired heaters are 40 and 50 gallons, the latter being a definite contender for

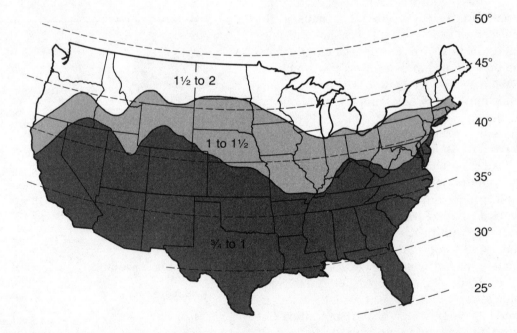

Figure 3–1: This map makes collector and storage sizing easy. The numbers shown are the square feet of collector area recommended per gallon of storage. For example, in the South, if you determine a need for 60 square feet of collector, you would need 60 to 80 gallons of storage (60 ÷ ¾; 60 ÷ 1). If you didn't want to buy a new storage tank, you could use the ratios the other way. In the far North, a 52-gallon electric water heater could be combined with 78 to 104 square feet of collector (52 × 1½; 52 × 2).

solar storage, and the former being possible for use with small collector areas. With flat plate systems, the yes-or-no decision on using the existing water heater depends on the amount of collector area needed to supply a large fraction (60 to 80 percent) of your hot water load. (In the single-pump drainback flat plate system, discussed in chapter 6, the size of the water heater is immaterial.) Here again the conservation connection becomes an important factor: Reduced load means reduced collector area, which means reduced storage tank volume, which makes it more likely that you can use the existing unit. Of course the dollar savings from that and from the reduced collector area are quite significant.

Another guideline you can use is the information in figure 3–1: This map shows recommended flat plate collector area/storage volume ratios for different parts of the United States and Canada. (Sizing guidelines for batch heaters will be discussed shortly.) As the map shows, in the south it's possible to combine ¾ to 1 square foot of collector area with 1 gallon of storage. This means that on sunny days each square foot of collector will be able to raise the temperature of at least 1 gallon of water to at least 120° to 130°F. Moving farther north, where there is generally less and less available solar energy, the collector area per gallon naturally increases until it reaches a ratio of 2 square feet per gallon.

You can, of course, turn the tables and see just how many square feet of collector area can be combined with your existing water heater. Then when you do the actual collector output and sizing calculation later in this chapter, you can see if the area allowed by the water heater is close to the area you calculate to supply at least 60 percent of your load. It is acceptable to oversize or undersize the collector a little (up to 10 percent relative to storage volume), but if you overdo it, you'll waste money on excess collector area, and you'll be putting too much heat into too small a tank, thereby wasting heat. Undersizing is OK except that it naturally limits the solar contribution to your water heating load. The calculation shown later in the chapter is aimed at finding an optimum economic collector size based on your load, your climate and other factors. But if you're budget conscious, you can minimize the system cost by using figure 3–1 to simply size the collectors to be compatible with the volume of your existing water heater.

Collector sizing may also be affected by physical limitations on available installation space, or by the module size of the collectors you build. You may calculate that you need 130 square feet of collector, but find that you can't fit that much into your chosen installation spot. Or, the collector you're working with is a 30-square-foot module. Four modules give you 120 square feet, but you can't, of course, work in one-third of a module (10 square feet). That's OK. Sizing guidelines are nothing if not flexible.

Another consideration on using the existing water heater is that you may be on the verge of retiring an old unit. The replacement you choose may need to be bigger than its predecessor to better receive the output of the collector area you settle on. This decision may also be affected by the type of system you choose. An antifreeze system, for example, uses a storage tank with a built-in heat exchanger and a built-in back-up electric heating element. You need to buy only this one tank to serve both the solar and water heater functions.

Bear in mind that these preliminary comparisons of hot water use, collector area and storage size are intended to give you a ballpark estimate of the size of the solar system you'll be working with. Flexibility in sizing has been stressed, but later in this chapter you'll learn how to compute the exact amount of heat produced by each square foot of collector to determine your *solar fraction* (the amount of energy supplied by the sun compared to the total amount of energy used to heat all your water). To get a higher solar fraction you can increase the size of the collector and go through the calculation again. But to use the calculation (which works for both batch and flat plate systems), you need to select a certain system type. You can, of course, make calculations on several systems that you could use in your application. To help you with your "which system?" decision, we've created the Joneses and the Smiths, similar families who end up choosing quite different systems. Both families have been thinking about solar water heating enough to make what is called a *solar site assessment*, and before we journey through their decision-making we'll explain more about this very important step in your planning.

Your Site, Your Sun

Once you know roughly how much collector area you need, you can venture outside to assess the amount of sunlight that falls on your house and lot. A solar site assessment should tell you at least three things: how much sunlight is available for hot water production, whether it's possible to mount your collectors on the roof or in the yard, and whether trees or buildings will shade the collectors and limit hot water production. But before you make a site assessment it's good to know a little about how sunshine reaches the earth's surface.

For thousands of years people believed that the earth was the center of the universe and that the sun—and everything else—revolved around it. This misconception was fostered because, from Earth, it *looks* like the sun marches across the sky, following a prescribed route. Early skywatchers, such as the builders of Stonehenge, observed that the sun's route predictably changes every season. They were right, but for the wrong reason. In the sixteenth century Nicolaus Copernicus theorized that the earth revolves around the sun, and today we of course know that the earth isn't the center of the universe. The sun's arc across the sky is lowest in winter and gradually climbs higher until the summer solstice (around June 21) when, year after year, it reaches its highest point. Then, as autumn approaches, the sun takes a lower and lower path across the sky until it "bottoms out" at the winter solstice (around December 21). The change in apparent height refers to *solar altitude*, which is commonly expressed as an angle of a given number of degrees. *Azimuth* refers to an angular measurement of the sun's compass position relative to a fixed reference direction. For example, if you're facing magnetic south (180 degrees on the compass) and the sun is found to be southeast (135 degrees), then the *azimuth angle* between the sun and the reference direction is 45 degrees (180 degrees to 135 degrees).

When studying the sun's path to determine the availability of solar energy, it's actually convenient to think of the sun as revolving around a fixed point (your home—the center of "your" solar system!). By understanding the sun's seasonal paths you can find the best direction to aim your solar collectors to get the most energy from the sun. By understanding how solar altitude changes with the seasons, you'll also ensure that shadows won't be cast on the collectors.

It took many years for the builders of Stonehenge to plot the sun's path across the

Photo 3–1: Here is a fusion reactor 93 million miles out in space, heating household water on earth to 120°F with little difficulty. This particular reactor is part of a public utility system that gives life but sends no monthly bills.

sky, but you can do it in minutes. Figures 3–3 through 3–10 are sun path charts that are relevant for different latitudes. These charts will help you visualize the seasonal paths of the sun. As you can see, the sun's height above the horizon varies depending on latitude. Find your latitude from the map in figure 3–11, then go to the appropriate sun path chart. The eight charts cover a broad expanse of North America, from south Texas (28 degrees north latitude) to the middle of Canada (56 degrees N.L.), in increments of 4 degrees. Being in between increments is no problem: If you live in East Wareham, Massachusetts (about 42 degrees N.L.), you can use the sun path chart for 40 degrees N.L. with no significant compromise on accuracy.

Photocopy the appropriate sun path chart and take it outside with you (is the sun shining?). Once outside, your first job is to find *true south*, which, it turns out, is not quite the same as magnetic south. For most locations, true south is a few degrees east or west of magnetic south. You can use the map in figure 3–11 to determine the extent of the variation in your area. The map shows that on an imaginary line that runs from the eastern side of Lake Michigan to the Atlantic Coast in northern Georgia there is no deviation. If you lived along that line, magnetic south would also be true south. Anywhere east of the line, true south is X degrees west of magnetic south, and anywhere west of that line, true south is X degrees east of magnetic south. All the squiggly lines in figure 3–11 tell you how many degrees of variation there are at your location. For example, in Philadelphia, true south is 8 degrees west of magnetic south. If you live in San Francisco, true south is 17 degrees east of magnetic south. Also, on any given day, when the sun reaches its highest

altitude (around noon), it is also in a true south position.

True south is the ideal direction for solar collectors to face. Your collectors will capture the maximum amount of solar energy if they're pointed in that direction, although they'll still be effective for water heating if they face as much as 30 degrees east or west of true south.

Once you know where true south is, you'll want to find the side of your house that's closest to that optimum bearing or *aspect*. Then you can look for the best place to install the collectors, using your rough estimate of the collector area you need. What you're looking for, in a nutshell, is an unshaded, southern exposure. Also, the potential collector site should be as close to your house as possible so there won't be excessively long pipe runs between the collectors and the storage tank. Both of these criteria can often be met by placing the collectors on your roof. A roof is commonly shade free, and it's about as close to your house as you can get without going inside. Also, most roofs are pitched steeply enough to give the collectors the proper tilt. The rule of thumb is that the tilt angle should be about equal to your latitude, plus or minus 10 degrees. For instance, if you live in Philadelphia at 40 degrees N.L., the collectors should ideally be tilted at 40 degrees, but a 30- to 50-degree tilt would be adequate. If your roof isn't already tilted to the proper angle, you can build special racks to hold the collectors, as described in a later chapter.

Up on the Roof

Does the roof face south? If not, don't despair. As photos 3–2 through 3–4 show, it's possible to mount the collectors so that they'll face south, even if the roof faces east or west. Is the roof area you're looking at big

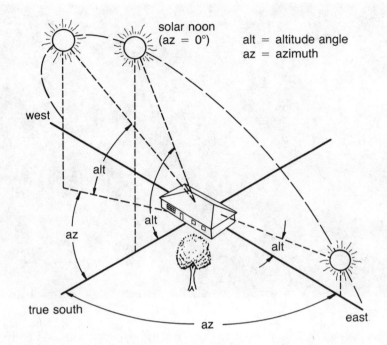

solar noon
(az = 0°)

alt = altitude angle
az = azimuth

west

alt

alt

az

alt

true south

east

az

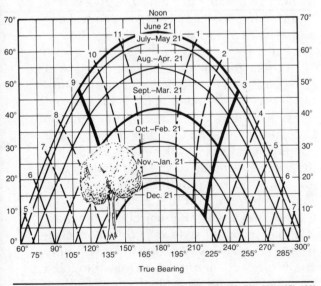

Figure 3–2: The drawing above visualizes a house as it relates to different solar altitude and azimuth angles and to solar noon, which is the sun's daily high point in the sky. In summer the sun's path is higher (altitude) and wider (azimuth) than in winter. Knowledge of the sun's seasonal paths will help you figure out whether or not shadows will fall across your planned collector site. This is where the sun path charts come in handy. The drawing at left suggests what a sun path chart could reveal if you crouched on the roof of the house in the top drawing. The tree blocks the sunlight between 8 and 10 o'clock in the morning through much of the fall, winter and spring months. This information might lead you to install the collectors someplace else. Ideally, collectors should be installed in a spot that's shade free from 9 A.M. to 3 P.M. (10 to 4 during daylight saving time).

enough to hold the collector area you've estimated? One nice thing about flat plate collectors (especially the kind you make yourself) is that they come in many sizes. If you need about 65 square feet of collector area, you can use three collectors that are 2 feet wide by 10 feet long (for a total of 60 square feet), or perhaps three collectors that are 3 feet wide by 8 feet long (for 72 square feet), and so on. With that kind of freedom you may have an easier time of fitting collectors onto your roof.

If the roof is big enough and faces the right direction, you'll want to make sure that it is not shaded most of the year, since you, of course, need hot water year-round. You must study tall objects like trees and buildings to make sure they won't cast shadows on the roof. This is where the sun path charts again come in handy.

Each chart has seven arching lines that are identified with different times of the year.

The highest arch represents the sun's path on June 21, the summer solstice, while the lowest arch represents the sun's low trek across the heavens on December 21, the shortest day of the year (winter solstice).

Sun Paths

Proper use of the sun path charts will give you a good idea if there's anything standing between your favorite collector site and the sun. Think of the June 21 line and the December 21 line on the chart as the top and bottom margins of a "solar energy zone." For the best year-round solar water heating performance, no trees, buildings or mountains should protrude into this zone (figure 3–2). If there are immovable objects in the way, you may have to locate your collectors elsewhere. (Trees can always be cut down—only as a last resort—but if you're forced to fell a tree, maybe you can replant a few saplings somewhere out of the way.) The sun path charts are most useful for making a general

(continued on page 50)

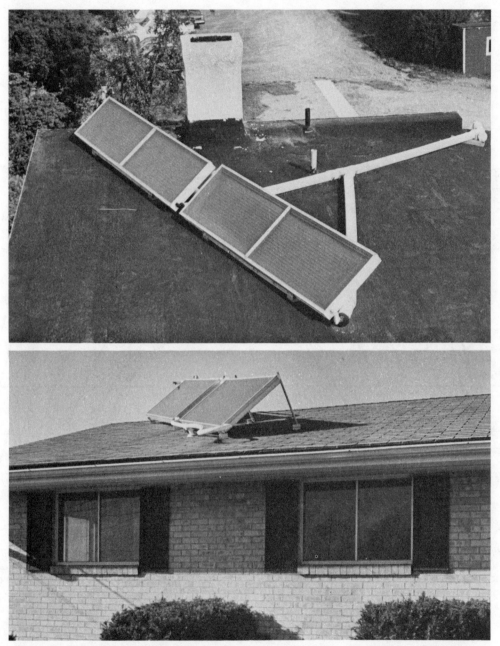

Photos 3–2, 3–3, 3–4: If the roof of a house doesn't face true south, or isn't slanted adequately, racks can compensate. The photo on the opposite page shows correction for tilt; the photos above show correction for tilt and azimuth on a flat roof, top, *and azimuth on a pitched roof,* bottom.

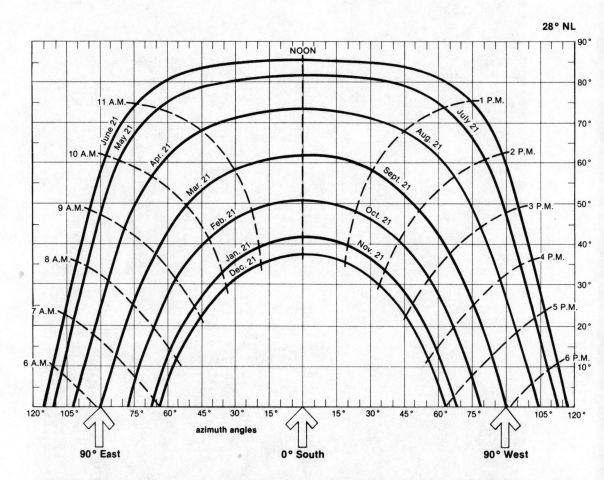

Figures 3–3 to 3–10: On the next several pages are eight sun path charts, each marked with a latitude number at the upper right-hand corner. Choose the chart whose latitude marking most closely corresponds to your site and you can see where the sun is in different seasons at different times of the day.

32° NL

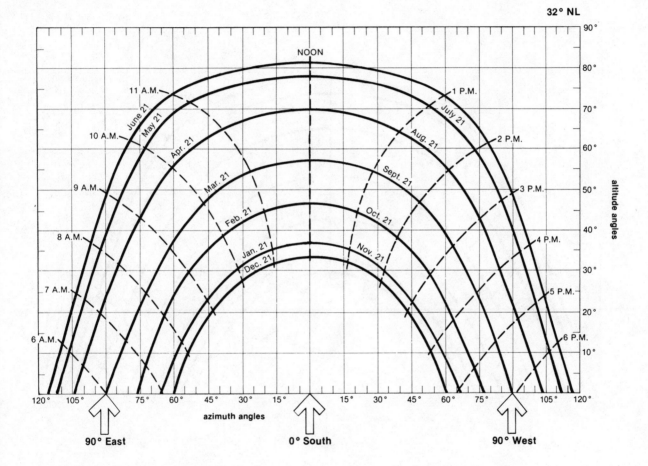

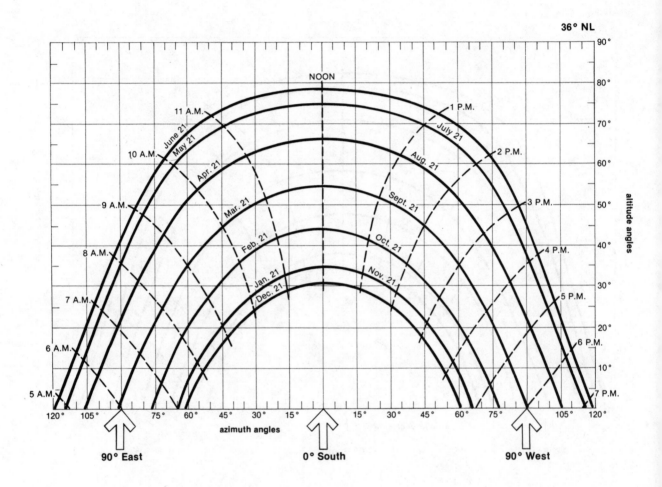

36° NL

NOON

11 A.M.   1 P.M.

June 21   July 21

May 21   Aug. 21

10 A.M.   2 P.M.

Apr. 21   Sept. 21

9 A.M.   3 P.M.

Mar. 21

Feb. 21   Oct. 21

8 A.M.   4 P.M.

Jan. 21   Nov. 21

Dec. 21

7 A.M.   5 P.M.

6 A.M.   6 P.M.

5 A.M.   7 P.M.

altitude angles

90°
80°
70°
60°
50°
40°
30°
20°
10°

120° 105° 90° 75° 60° 45° 30° 15°   15° 30° 45° 60° 75° 105° 120°

azimuth angles

90° East   0° South   90° West

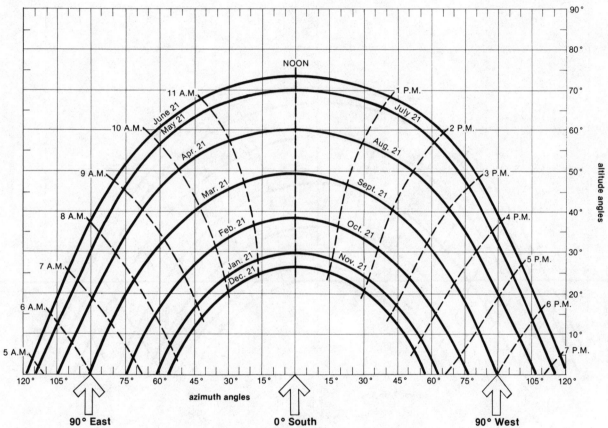

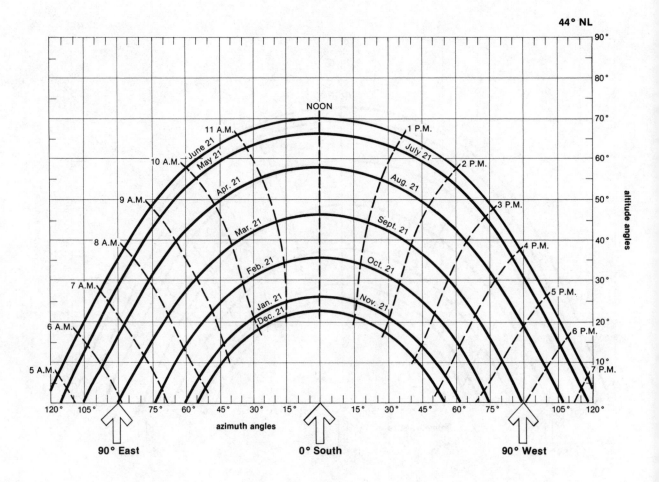

44° NL

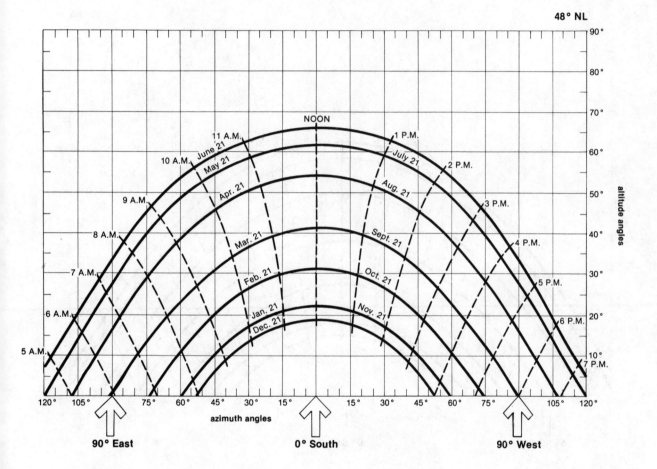

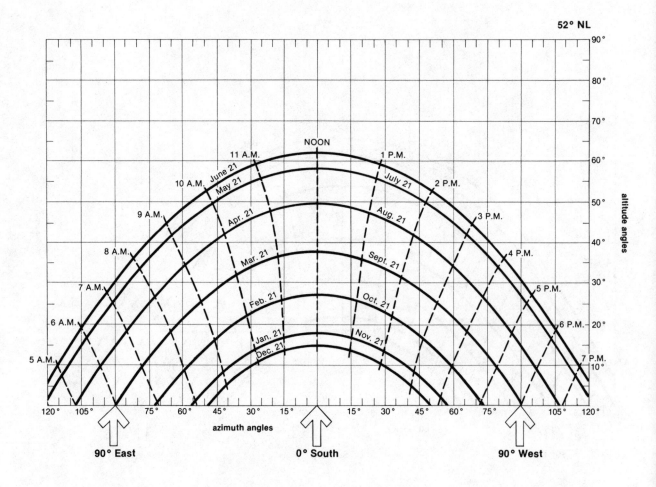

52° NL

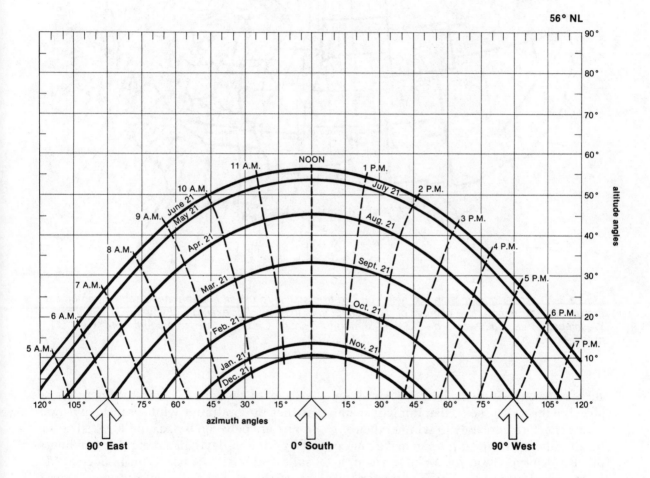

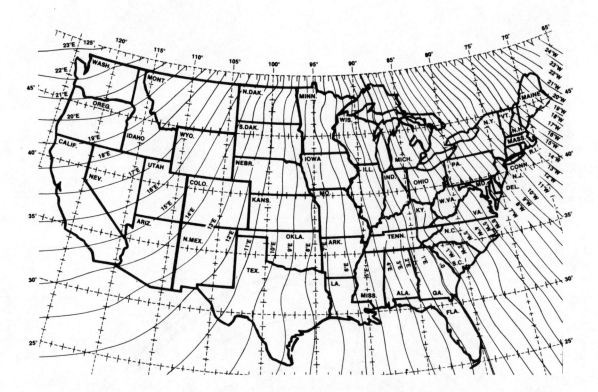

Figure 3–11: This isogonic map will help you find true south. To use it, find magnetic south, then adjust the reading the appropriate number of degrees for your location. For example, if you lived in Seattle, Washington, true south would be 22° east of magnetic south. On Cape Cod, true south is 14° west of magnetic south.

survey of your southern field of view. If an object to the south *seems* like it might shade your collectors (probably in winter when solar altitudes are lowest), you can make a more precise determination by making a simple sighting tool using a common protractor and a piece of straight wire. Figure 3–12 shows that a small hole is drilled into the center of the protractor's straight side, and that the wire is bent so as to hook into the hole and still swing freely.

Say, for example, that you want to see if a tall tree will shade your collectors (located at 44 degrees N.L.) on December 21, the short-

est day of the year. The first thing to know is that the optimum winter solar collection hours are between 9 A.M. and 3 P.M. (During the period of daylight saving time the hours shift to 10 A.M. to 4 P.M.) The 44-degree N.L. chart shows that the solar altitude at 9 or 3 o'clock is 12 degrees, when the sun's azimuth angle is 40 degrees. The highest altitude for the day is about 23 degrees. You can take that information out to your roof, or wherever it is you're contemplating the installation.

The sighting tool is used to "shoot" the treetop as shown in figure 3–12. When you use it in the planned collector location, the swinging wire will indicate the angular altitude of the tree. If this angle exceeds the angles indicated on the sun path chart, there will be a shading problem. To see if it's a long-term problem, you can check the tree's altitude angle against those indicated for the January/November and February/October sun paths.

A more precise method for locating shade-makers is by using some simple mathematics. Figure 3–13 explains how you can use trigonometry to get a fix on where shadows will fall on a roof or vertical wall or in your yard. By making these calculations you can accurately prove or disprove any suspected shadow-makers. In most cases, however, your unaided eyes and common sense will tell you if there's something in the way that will prevent sunshine from reaching the collectors.

If nearby objects rule out installing the collectors on the roof, you still have the option of placing them on the lawn or on a south-facing vertical wall. The latter option may well be eliminated because a shaded roof often means a shaded wall, but take a look anyway. Ideally, a lawn site should be as close to the house as possible, and if more than one site is available, choose the one that's also closest to where the storage tank is or will be.

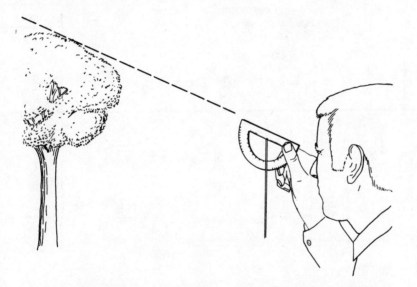

Figure 3–12: Here's a simple, but accurate, sighting tool you can make with an ordinary protractor and a piece of straight wire that hangs in a hole drilled at the middle of the straight edge. Stand as close as possible to the surface where you want to install your collector. Site along the straight edge to the top of any object you suspect might be a shadow-maker. Note the altitude angle indicated by the wire. You can take that information to the sun path chart to see if in fact the thing will block the sun during the optimum collection hours (9 A.M. to 3 P.M.).

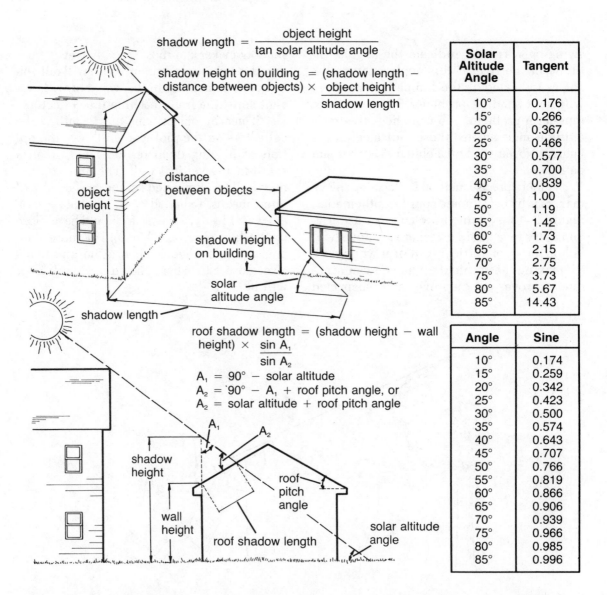

$$\text{shadow length} = \frac{\text{object height}}{\text{tan solar altitude angle}}$$

shadow height on building = (shadow length − distance between objects) × $\dfrac{\text{object height}}{\text{shadow length}}$

Solar Altitude Angle	Tangent
10°	0.176
15°	0.266
20°	0.367
25°	0.466
30°	0.577
35°	0.700
40°	0.839
45°	1.00
50°	1.19
55°	1.42
60°	1.73
65°	2.15
70°	2.75
75°	3.73
80°	5.67
85°	14.43

roof shadow length = (shadow height − wall height) × $\dfrac{\sin A_1}{\sin A_2}$

$A_1 = 90° -$ solar altitude
$A_2 = 90° - A_1 +$ roof pitch angle, or
$A_2 =$ solar altitude + roof pitch angle

Angle	Sine
10°	0.174
15°	0.259
20°	0.342
25°	0.423
30°	0.500
35°	0.574
40°	0.643
45°	0.707
50°	0.766
55°	0.819
60°	0.866
65°	0.906
70°	0.939
75°	0.966
80°	0.985
85°	0.996

Figure 3–13: The top illustration and calculation provide everything you need to find the height of a shadow on a vertical wall. To find the length of a shadow cast across a pitched roof, use the bottom illustration and calculation. For example, if you found in the top calculation that a shadow height is 15 feet, and the vertical wall height of the building is 10 feet, the roof pitch angle is 20° and the solar altitude is 30°, then:

$$A_1 = 90° - 30° = 60°; \ \sin 60° = 0.866$$
$$A_2 = 90° - 60° + 20° = 50°; \ \sin 50° = 0.766$$

$$\textit{roof shadow length} = (15 \text{ ft.} - 10 \text{ ft.}) \times \frac{0.866}{0.766} = 5.65 \text{ ft.}$$

Making a Decision

In the preceding chapter, several decision-making variables were introduced: the cost of the system, whether you want to build all or part of the system yourself, the physical layout of your house and yard, the climate and ensuing freeze protection requirements, the structural integrity of your roof, aesthetics, ease of operation and maintenance and hot water consumption.

To help you size and plan your system, let's consider the homes of our two hypothetical families, the Smiths and the Joneses. They live near Hartford, Connecticut, in a climate zone that figure 3–1 indicates requires at least a foot and a half of collector area for every gallon of storage. To save money, the Joneses want to use their existing 65-gallon water heater as the solar storage tank, so Mr. Jones figures he'll need around 97½ square feet of collector area. There are many ways for Mr. Jones to get that 97½ square feet. In theory, he can build two collectors that are each 5 feet wide by 10 feet long, thus providing 100 square feet of collector surface, but such unwieldy collectors would be too difficult to haul up onto the roof. In the back of his mind Mr. Jones considers building three 4 by 8-foot collectors, with a total of 96 square feet of collector area, certainly close enough to the recommended 97½ square feet.

Now Mr. Jones mulls over the variables.

He first compares the costs of mounting the collectors on the roof versus installing them in the yard. He believes, incorrectly, that it will cost the same either way, so cost isn't yet a deciding factor. He would like to try building the system himself, but that doesn't rule out roof- or lawn-mounted collectors, since an experienced do-it-yourselfer could install collectors in either place.

The physical layout of the house, however, immediately affects his decision. Mr. Jones would like to build a passive flat plate system (thermosiphon) because its simplicity appeals to him. He knows, however, that in a thermosiphon system the storage tank must be above the collectors. There's the rub. He can't mount the collectors on the roof because there's no way to get the tank higher than the top of the collectors. On the other hand, if he were to install the collectors on his sloping lawn, he could put the storage tank up near the ceiling in the first-floor laundry room, which would place the tank about 2 or 3 feet above the collectors. So far, so good. Since a roof mount would mean installing a more expensive, active system, Mr. Jones figures he can probably count on saving between $500 and $1000 by sticking with the thermosiphon option. Thus the cost difference between the roof-mounted collectors and the lawn collectors turns out to be an important variable.

The fifth variable, the structural integrity of the roof, doesn't concern Mr. Jones, since his roof is strong and can easily support the weight of the flat plates. Just about every roof can handle the well-distributed weight of flat plate collectors. Nevertheless, you should check with an expert to make sure that your roof really is strong enough. You may also be required by your town, city or county to get a building permit for the collector addition.

Mr. Jones now addresses the question of having a good-looking installation. It occurs to him that if he installs the flat plate collectors on the lawn, which he must do if he wants a thermosiphon system, he'll have to eliminate his wife's flower bed. Mr. Jones knows he's free to do just about anything under the sun, except go near Mrs. Jones's flowers. The collectors would turn the bed into a sunless rock garden, and Mrs. Jones is quick to react: She'd rather have them out of sight of everything except the sun. Mr. Jones realizes that he'll probably have to install an active system with the collectors on the roof, even though that would be more expensive than a simpler, passive system. It's also true that the passive system wouldn't have had any automatic freeze protection, a must in the Hartford winter. To prevent freezing damage the collectors would have to be drained every evening or through the whole winter (which would greatly reduce the annual productivity of the system). The active system, on the other hand, is automated with a supposedly foolproof freeze protection system. The pump and thermostatic controller could require maintenance over time, but this system does have the advantage of year-round operation. Mr. Jones had thought a little about installing a passive batch system, but, again, the flower bed won out.

Finally, Mr. Jones considers the last solar variable, which is hot water consumption. Either a batch heater or a passive or active flat plate system could provide the Joneses with lots of hot water. By minimizing their water heating load with conservation measures, this variable became less important than others. The larger the load, however, the more an active flat plate system is likely to be needed, especially during the winter months.

Having weighed all the variables, Mr. Jones concludes that he should mount the collectors on the roof and that the system should be active, and he begins the next stage of planning the collector location, the pipe runs, where the controls will be installed, and other details of the installation.

When you're heading for your own decision on a system, you'll probably start with the most influential variable—cost. Often your financial situation will dictate the type of system you'll eventually install. Solar water heating systems can cost as little as $300 or $400 (for a single-tank batch water heater that you build yourself) or as much as $4000 (for a fully automated, state-of-the-art, commercially built and installed flat plate system). Once you know how much you want to spend, you're well on your way to deciding which system is best for your home.

In most cases, you can build a good thermosiphon system with flat plate collectors for about $1000. Active systems generally cost more than passive systems because there's the added expense of pumps and a thermostatic controller. An active system with flat plate collectors will usually cost a do-it-yourselfer between $1000 and $2000, depending on the size of the system. It's important to know that the wide price differences between, say, a batch system and an active flat plate system don't mean that the latter will automatically produce three or four times more hot water than the former. A properly sized batch system, at one-fourth to one-fifth the cost of a flat plate system, can have a hot water output of one-half to two-thirds that of a comparable flat plate system.

The Smiths, our second fictional family, were intrigued by that possibility when they began their planning. The Smiths' site assessment showed that none of their house's roof pitches were oriented within 30 degrees east or west of true south. Adding to that difficulty, a neighbor's house casts a long shadow in winter, which ruled out installing collectors on the roof. It turned out that only certain areas of the lawn got year-round sunshine. The Smiths were thus looking at a "lawn job," not the lawn mower kind, but the where-to-put-the-collectors kind.

In considering the cost variable, Mr. Smith knows he can spend around $500, so it becomes an easy decision to build a batch collector for a sunny space on the lawn that's right next to the house. With the ground mount there aren't, of course, any structural problems to consider, as with a roof mount, and the design that Mr. Smith envisions for the collector is one that will blend with the look of the existing house. (Nor is the fate of any bed of flowers in question.)

The Smiths know that a batch system isn't quite as easy to use as an active flat plate system. In very cold climates it must usually be drained in winter because of the lack of automatic freeze protection. In Hartford, though, the collector could be protected with insulated hatch covers that are closed at night, a chore that could be handled by one of the teenagers. This allows year-round use of the system, and the Smiths will have a simple, reliable system with no moving parts, no electrical controls.

How Much Solar Energy Can You Get?

The Smiths and the Joneses have both reached decisions that reflect their different situations and needs, and they're ready to take the solar plunge. But before they, or you or anyone, start planning the actual installation, the value of that installation, in terms of the solar contribution to the water heating load, can be predicted with a little number work.

What follows is a simple calculation system that allows you to make that prediction with reasonable accuracy so you can determine the annual dollar value of the energy collected by the solar system. Then you can see how long it will take for the accrued solar "dividends" to exceed the cost of the installation. This is called the *payback period*.

Finding Your Water Heating Load

The first step in this calculation (worksheet 3–1) is to refer again to your household's daily hot water gallonage figures, which you computed earlier. Next, you can calculate how many Btu's are consumed to heat all those gallons to determine the annual water heating load. Finally, you can work in your cost for energy to find out the price for a year's worth of hot water. Then it's just a few more steps to find the estimated "heat gain" of the solar water heating system you're planning to build or buy (worksheet 3–2). In this part of the calculation there are a number of factors to work in about climate, collector orientation, system type and others, all of which are listed in tables for easy reference. In the following pages you'll be literally led by the calculator through this estimation, so if you feel a little intimidated by an equation, fear not; you're getting a thoroughly guided tour.

Worksheet 3–1 will help you compute your household's annual Btu consumption for water heating. (If you photocopy it, you can use it again and again.) On lines 1 through 6 of the worksheet fill in your home's daily hot water consumption for the various uses listed. On line 7 total up the daily gallonages, and on line 8 divide that result by the number of people in your household to get gallons per person per day.

Next you find the *temperature rise*, which is the difference in degrees Fahrenheit between water entering and exiting the water heater. In most regions the temperature of water entering the heater is about equal to the average outside air temperature. This is usually true, except that the average water temperature hardly ever drops below 40°F, while in very cold climates the average air temperature can get much lower. To find your area's average water inlet temperature see ta-

Worksheet 3–1   Estimating Your Domestic Hot Water (DHW) Energy Requirements

Calculation	Line No.	How Obtained	Examples		Your Household
			Average Use (family of 4)	Conservative Use (family of 4)	
Gallonage for: Clothes washing	1	see table 3–1	automatic washing machine; hot cycle; 1 load every 2 days — 11 gal./day	automatic washing machine; warm or cold cycle; 1 load every 2 days — 4 gal./day	_____
Dish washing	2	see table 3–1	automatic dishwasher, 1 load per day — 15 gal./day	dish washing by hand — 4 gal./day	_____
Food preparation	3	see table 3–1	3 gal./day	2 gal./day	_____
Household cleaning	4	see table 3–1	2 gal./day	2 gal./day	_____
Hand and face washing and shaving	5	see table 3–1	10 gal./day	flow controls in faucets — 8 gal./day	_____
Baths and showers	6	see table 3–1	regular shower head; shower or bath every other day — 40 gal./day	low-flow shower head; shower every other day; quicker showers; no baths — 20 gal./day	_____
Daily Total Gallonage	7	add 1 through 6	81 gal./day	40 gal./day	_____
Daily Total Gallonage/Person	8	7 / no. of people in household	$\frac{81}{4} = 20.25$ gal./person/day	$\frac{40}{4} = 10$ gal./person/day	_____
Cold Water Inlet Temperature (°F)	9	see text	50°F	50°F	_____
Hot Water Delivery Temperature (°F)	10	see text	140°F	110°F	_____
Temperature Rise	11	subtract 9 from 10	$140° - 50° = 90°F$	$110° - 50° = 60°F$	_____
Energy Use from Hot Water Consumption	12	$\frac{7 \times 11 \times 8.33 \times 365}{1,000,000}$	$\frac{81 \times 90 \times 8.33 \times 365}{1,000,000} = 22.17$ MBtu/yr.	$\frac{40 \times 60 \times 8.33 \times 365}{1,000,000} = 7.30$ MBtu/yr.	_____
Standby Losses	13	see text	$0.20 \times 22.17 = 4.43$ MBtu/yr.	$0.05 \times 7.30 = 0.37$ MBtu/yr.	_____
Total Annual DHW Energy Use	14	12 + 13	$22.17 + 4.43 = 26.6$ MBtu/yr.	$7.30 + 0.37 = 7.67$ MBtu/yr.	_____
Total Annual DHW Energy Use/Person	15	14 / no. of people in household	$\frac{26.6}{4} = 6.65$ MBtu/person/yr.	$\frac{7.67}{4} = 1.92$ MBtu/person/yr.	_____
Total DHW Energy Use/Person/Day	16	$\frac{15}{365}$	$\frac{6.65}{365} = 0.01821$ MBtu/person/day $= 18,210$ Btu/person/day	$\frac{1.92}{365} = 0.052$ MBtu/person/day $= 5250$ Btu/person/day	_____
Annual DHW Energy Cost	17	14 × energy cost from table 3–5	$26.6 \times \$20.51$/MBtu $= \$546$/yr. (electricity @ 7¢/KWH)	$7.67 \times \$20.51$/MBtu $= \$157$/yr. (electricity @ 7¢/KWH)	_____

SOURCE: Adapted from *Solarizing Your Present Home*, Joe Carter, ed. (Emmaus, Pa.: Rodale Press, 1981).

ble 3–2, column A, which lists average air temperatures for over 170 locations. Write in the inlet water temperature on line 9 of the worksheet. (If the number listed for your location is below 40°F, use 40°F on line 9.) The next step is to measure your water heater's outlet temperature. During a time of the day when no one is showering or washing clothes, open the hot water faucet nearest the water heater. Allow hot water to fill and run over a cup while you put in a Fahrenheit thermometer. After two or three minutes the thermometer will show the heater's outlet temperature, which you should record on line 10 of the worksheet. Subtract the inlet temperature (line 9) from the outlet temperature (line 10) and the result will be the temperature rise. (If, for example, your average inlet water temperature is 50°F, and the outlet temperature is 110°F, then the temperature rise is 60°F.) Enter the result on line 11 of the worksheet.

Next, multiply your total daily hot water gallonage (line 7) times the temperature rise (line 11) times 8.33 (the weight of a gallon of water) times 365. Divide the product by 1,000,000 and enter the result on line 12.

Now add your water heating system's efficiency into the equation. If you followed the advice in chapter 1 and wrapped 6 inches of fiberglass insulation around your water heater, covered the hot water pipes with R-4 insulation, and set back the water heater's thermostat to 110° to 120°F, multiply line 12 by 0.05, and enter the result on line 13. If you haven't insulated the heater and pipe runs, and the heater still produces water at temperatures greater than 120°F, multiply line 12 by 0.20 to find line 13.

Now add lines 12 and 13 and enter the result on line 14. This sum is your home's yearly Btu consumption (in MBtu) for water heating. This is the important number that you'll compare with the annual solar energy gain of the system you're planning to install. (It's not assumed that you've actually made the choice already. Read on through this book and find out more about the various types of systems. Then you can come back with your choice, or even compare the performance of different systems.)

Lines 15 and 16 are simply ways to calculate your per-person water heating load per year (line 14 is divided by the number of people in the household) and per day (line 15 is divided by 365). What may be more interesting, though, is what all this hot water is costing you. In line 17 you multiply the total load from line 14 times your energy cost as drawn from table 3–5. To use the table, find the unit cost of the energy that's powering your conventional water heater. Electricity is usually sold by the kilowatt-hour. Gas is often in units of therms (100 cubic feet). On your gas or electric bill there should be a listing of the unit price. For example, these days gas might sell for about 60¢ a therm, while electricity costs about 7¢ a kilowatt-hour. But you should also take into account all the little taxes and surcharges that can significantly boost the unit cost. The best way to figure your unit cost is to divide the total bill by the number of units you're being billed for.

Next, multiply that unit cost times the appropriate energy cost factor from the table. The energy cost factors convert the cost of a single unit into a price per million Btu's. These factors also take into account the different efficiencies at which these energy units are utilized. If you have an electric water heater, you'd multiply the cost of a kilowatt-hour (for example, $.07) times the energy cost factor for electricity (293), to find that $20.51 is the price for a million Btu's. If you use gas, you'd multiply the cost of a therm (for example, $.60) times the energy cost factor for natural gas (14.29) to determine that $8.57 is the price for a million Btu's.

If you've made the conservation improvements discussed in chapter 1, and you're still paying over $150 a year for hot water, you're probably a good candidate for a solar water heater. To find out, you can complete the next worksheet.

Finding the Solar Fraction

Now that you know how much energy goes into your water every year, the next step is to estimate how many of those Btu's can be solar Btu's. Worksheet 3–2 will help you do that, and to show you exactly how it's done, the worksheet includes a sample calculation based on the decisions made by one of our fictional friends, the Joneses.

Collector Area and Gross Solar Gain

On line 1 of the worksheet, enter your system's total square feet of collector area. (If you're building the basic batch heater according to the instructions found in the next chapter, the collector area will be about 31.5 square feet.) How do you know what that number should be? You don't, but for starters work with 15 square feet per person in the household. If the result at the end of the calculation is too high or too low, you can adjust the square footage accordingly.

Line 2 is where you enter the amount of solar energy that falls every year on a square foot of horizontal surface in your area. This information is found in table 3–2, column B, near the end of this chapter. In Hartford, Connecticut, for example, the number is about 457,000 or 457 KBtu/ft²/yr (KBtu = Btu ÷ 1000).

Collector Tilt and Orientation Factor

Because solar collectors are usually tilted toward the sun's path, they aren't true "horizontal surfaces." This has to be accounted for in the calculation. Along with variations in tilt there are also variations in collector orientation, since of course all collectors can't be oriented exactly to true south. When collectors do face X number of degrees east or west of true south, some efficiency is lost. Efficiency is also affected by the tilt angle relative to your latitude. A collector will catch the most sunlight year-round if its tilt angle is equal to the latitude at the site. In other words, a collector in Columbus, Ohio (40 degrees N.L.), should have a 40-degree tilt, while a collector in Fairbanks, Alaska (64° N.L.),

Photo 3–5: Tilt angles are found in an instant with an inclinometer. It is to altitude angles what the compass is to azimuth.

Worksheet 3–2   Solar Domestic Hot Water (DHW) System Performance Calculation

Calculation	Line No.	How Obtained	Example	Your Home Case A	Case B
Collector area (clear aperture)	1	your own measurements	34.0 ft^2		
Annual total solar energy received on horizontal surface	2	see table 3–2, column B	423 KBtu/ft^2/yr.		
Adjustment factor for collector tilt and orientation	3	see table 3–3	1.10		
Annual total solar energy received on collector	4	$\dfrac{1 \times 2 \times 3}{1000}$	$\dfrac{423 \times 1.10 \times 34.0}{1000} = 15.82$ MBtu/yr.		
Outer glazing transmittance	5	see table 3–4	0.85		
Inner glazing transmittance	6	see table 3–4	0.71		
Glazing averaging factor	7	see table 3–4	0.90		
System heat gain factor	8	$5 \times 6 \times 7$	$0.85 \times 0.71 \times 0.90 = 0.54$		
Average annual temperature	9	see table 3–2, column A	51.4°F		
Collector loss factor	10	see table 3–5	0.50		
Glazing loss factor	11	see table 3–5	0.80		
Absorber loss factor	12	see table 3–5	1.0		
Heat exchanger loss factor	13	see table 3–5	1.0		
System heat loss factor	14	$(80 - 9) \times 10 \times 11 \times 12 \times 13 \times 0.027$	$(80 - 51.4)(0.5)(0.8)(1)(1) \times (0.027) = 0.31$		
System efficiency	15	$8 - 14$	$0.54 - 0.31 = 0.23$		
DHW load (including standby losses)	16	see worksheet 3–1, line 14	7.67 MBtu/yr.		
Gross solar fraction	17	$\dfrac{4 \times 15}{16}$	$\dfrac{15.82 \times 0.23}{7.67} = 0.47$		
Net solar fraction	18	see table 3–6	0.47		
Annual energy savings	19	18×16	$0.47 \times 7.67 = 3.60$ MBtu/yr.		
Annual dollar savings	20	$18 \times$ worksheet 3–1, line 17	$0.47 \times \$157 = \$73.79$/yr.		

SOURCE: Adapted from *Solarizing Your Present Home*, Joe Carter, ed. (Emmaus, Pa.: Rodale Press, 1981).

should be tilted at about 64 degrees, and so on. (Your latitude can be found in figure 3–11.) Of course, a variation of 5 degrees either way won't make a great difference in overall performance, though a 10-degree variation would be significant. If you live at 40 degrees N.L. and your roof is pitched at a 30-degree angle, a simple rack system (described in chapter 5) will get you the extra 10 degrees. What's your roof angle? You can figure it out using the information in figure 3–14 or by

getting a nifty little tool called an inclinometer (photo 3–5).

While you may have a lot of control over tilt, it's not so easy to point the collectors toward true south. For example, perhaps your roof faces 15 degrees west of true south. It may well be less expensive to mount the collectors directly onto the roof and lose some efficiency than it is to build or buy special adjustable mounting brackets that will point the collectors toward true south. Solar collectors can face as much as 45 degrees east or west of true south and still produce some hot water, but deviations of more than 30 degrees result in a dramatic reduction in output. So if your mounting plane is way off true south and you don't want to see your collectors "cocked" at an odd angle relative to your house, you can increase collector area (by one module) to make up for the reduced output.

If your collector installation is less than "perfect" (vis-à-vis tilt and orientation), you can use table 3–3, which has adjustment factors for collector tilt and orientation. This factor is inserted on line 3 of the worksheet.

Line 4 of the worksheet is where you multiply together the three previous lines to find the total annual amount of solar energy that will be received by your collectors. This number will be on the order of several MBtu.

System Efficiency Factors

Now you can take into consideration the overall efficiency of the system you're planning. By going through this part of the calculation you'll also get a better idea of the kind of system that's best for your house, and the best materials to use in building the system.

Start with the efficiency of the collector's glazing. Figure 3–15 is a map of North America showing the two climate zones that determine the required number of glazing layers. It's simple to use: If you live in Zone II, the South, you'll need only one layer of glazing. If you live in Zone I, the North, you'll need two layers of glazing to hold in the col-

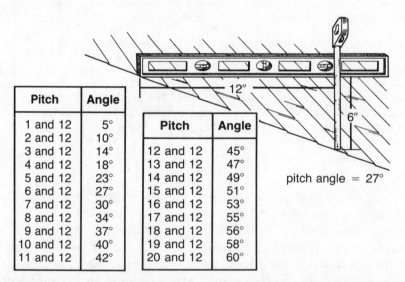

Pitch	Angle
1 and 12	5°
2 and 12	10°
3 and 12	14°
4 and 12	18°
5 and 12	23°
6 and 12	27°
7 and 12	30°
8 and 12	34°
9 and 12	37°
10 and 12	40°
11 and 12	42°

Pitch	Angle
12 and 12	45°
13 and 12	47°
14 and 12	49°
15 and 12	51°
16 and 12	53°
17 and 12	55°
18 and 12	56°
19 and 12	58°
20 and 12	60°

pitch angle = 27°

Figure 3–14: The angle of a roof can be found by placing one end of a level on the pitch and holding it horizontally. Twelve inches out from the end of the level, measure down to the roof with a vertically held tape. If, for example, you find there are 6 inches of rise per foot of run, you have a 6- and -12 pitch, which corresponds to an angle of 27°. (By the way, when working on a pitch steeper than 20°, scaffolding or a platform should be used for safety.)

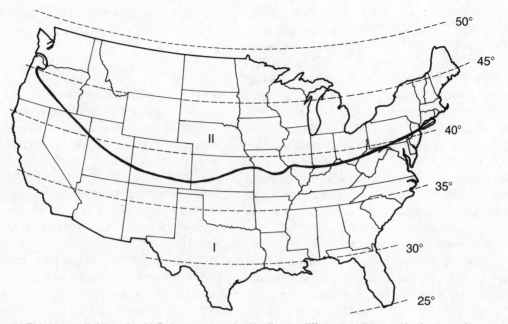

Figure 3–15: This map divides the U.S. into zones that indicate different collector glazing requirements. If you live in Zone I, flat collector plates need two layers and batch collectors need three. In Zone II, flat plates need just one layer, while batch collectors need two.

lector's heat. There is nowhere that you should use three layers for a flat plate unit, though three glazings are often used on a batch collector.

Table 3–4 lists glazing transmittance factors for various kinds of collector glazings (and explains the differences between them). *Transmittance* is the amount of solar energy passed by a layer of glazing relative to the amount reaching the outer surface. The best glazing material for the price is low-iron glass, which has a transmittance of 0.91 (91%). Two layers of this glass would have a transmittance of about 0.82 (0.91 × 0.91). For now, if you're not sure what type of glazing material you'll eventually buy, use the efficiency factor for low-iron glass. If you go shopping for glazing and wind up with something else, you can always redo this calcu-

lation using the actual transmittance of your glazing.

On line 5 of the worksheet enter the transmittance of your single outer glazing. If you're using a second, inner layer, enter that transmittance on line 6. Next, on line 7, enter the glazing averaging factor for your collectors, which is at the bottom of table 3–4. For collectors with a single layer of glazing, the glazing averaging factor is 0.93. For collectors with a double layer of glazing, the glazing averaging factor is 0.90. Now multiply lines 5, 6 and 7 together to find the system's overall heat gain factor, and enter it on line 8.

On line 9 enter your area's average annual ambient temperature (this information is found in table 3–2, column A).

Table 3–5 lists the heat loss factors for both batch and flat-plate collectors. Mark your collector's loss factor on line 10. The glazing loss factor, found in table 3–5, is entered on line 11. The absorber loss factor, also found in table 3–5, is entered on line 12. It's recommended that you spend a little bit more to paint your absorber plates with a specially formulated *selective coating* rather than ordinary flat black paint, which neither absorbs solar energy nor retains heat as well as a selective coating. (This will be discussed more thoroughly in chapter 5.) If you are going to cover your absorber plates with a recommended selective coating, mark "0.90" on line 12. If you're going to use a standard flat black paint, mark "1.0" on line 12.

Line 13 is where you enter the heat exchanger loss factor if the system you select is a closed-loop system that uses a heat exchanger. That would include the drainback and antifreeze active flat plate systems. If you're planning to use one of these systems, enter "1.0" on line 13. If you're going to use one of the other systems, enter "1.10" on line 13 (see table 3–5).

To find line 14 on the worksheet you must do a little math. First, subtract line 9, the ambient air temperature, from 80, which is the "balance temperature" for solar water heating systems. (Collectors warmer than 80°F start losing appreciable amounts of heat back to the atmosphere.) Multiply the difference times line 10, the collector loss factor. Then multiply the result times line 11, the glazing loss factor, and then multiply that product times line 12, the absorber loss factor. Then multiply that times line 13, the exchanger loss factor. Finally, multiply that product by 0.027. This final factor is a conversion factor that changes temperatures and individual heat loss factors into an overall heat loss factor (take our word for it). Line 14 represents your system's total heat loss factor. Line 15, the system's overall efficiency, is found by subtracting line 14 from line 8, the system heat gain factor.

Gross and Net Solar Fractions

Now enter your home's annual Btu consumption (calculated in worksheet 3–1) on line 16. You can now find your system's gross solar fraction by first multiplying line 4 (the total annual amount of sunlight falling on your collectors) times line 15 (the system's overall efficiency). Then divide this product by line 16; enter the gross solar fraction on line 17.

A *solar fraction* is the solar energy input to a total load, expressed as a fraction or percentage. Of course, the higher the solar fraction, the better, although there are economic limits to attaining very high solar fractions (over 85 percent). Table 3–6 gives you the conversions from the gross solar fraction to the net solar fraction, which you enter on line 18. For example, if your system's gross solar fraction is 0.75, then the net solar fraction would be 0.64. Your domestic water heating system would get 64 percent of its energy from the sun.

Now that you know the net solar fraction of the system you're planning, you can estimate your system's annual energy savings in terms of Btu's. Multiply line 18 times line 16 (your home's annual Btu consumption for water heating). The product is entered on line 19 as the annual energy savings (in MBtu) of the system you're planning.

To find how much money your planned solar water heater will save per year (at the current fuel cost), multiply line 18 (the net solar fraction) times line 17 from worksheet 3-1 (your annual DHW energy cost). To make a determination of the "simple" payback period of a solar heating system (which doesn't account for inflation, the cost of money, rising costs for conventional energy) simply divide the projected system cost by line 20 (annual dollar savings). The more you pay for energy (electricity is the most expensive per MBtu), the shorter your payback is likely to be. Also, it's important to note that until the end of 1984 there is a 40 percent solar energy income tax credit in the United States. This means that a $1000 system, for instance, will ultimately cost you only $600 because you deduct 40 percent of the cost of the system from your tax bill. It's not just a deduction; it's a true credit, and a healthy discount on the cost of solar heat.

Follow the same procedure to calculate the payback period of the system you're planning. When you do this you'll also get a good idea how you can "juggle" things around to get a payback period you'll be happy with. By now, for example, you understand the relationship between collector area and energy gain. If you want, you can calculate the costs and payback periods of several different systems, each slightly larger than the next; then choose the size system that seems the most financially attractive. Remember, the recommended flat plate collector area can be as much as 10 percent below your calculated need and still produce enough hot water for your household's needs. So if the recommended flat plate area is 100 square feet, your home can probably get by nicely with 90 square feet. Since commercially installed flat plate collectors can cost as much as $20 per square foot, it is well worth your while to calculate whether extra collector area will yield enough energy to make the added investment worthwhile.

TABLE 3–2

Temperature and Sunshine Conditions in the United States and Canada

Avg. Solar Energy Received on Horizontal Surfaces Annual Total (KBtu/ft²/yr.)

State or Province	Location	Ambient Temp. (°F) Annual Avg. A	B
Alabama	Birmingham	64.1	534
	Montgomery	65.0	543
Alaska	Fairbanks	40.0	321
Arizona	Flagstaff	45.6	691
	Phoenix	69.0	711
	Tucson	67.7	697
	Yuma	72.5	682
Arkansas	Fort Smith	61.8	519
	Little Rock	61.7	526
California, northern	Eureka	52.3	422
	Red Bluff	63.5	552
	Sacramento	60.4	587
	San Francisco	56.9	533
California, southern	Bakersfield	65.1	784
	Fresno	63.0	615
	Los Angeles Airport	no data	602
	Los Angeles City	61.9	614
	San Diego	63.2	548
	Santa Maria	57.0	657
Colorado	Denver	49.5	601
	Grand Junction	52.5	621
	Pueblo	52.7	625
Connecticut	Hartford	49.8	457
	New Haven	50.2	474
Florida	Jacksonville	69.5	544
	Miami	76.2	606
	Pensacola	68.4	560
	Tallahassee	68.0	604
	Tampa	72.2	618

(continued on next page)

SOURCE: Adapted from *Solarizing Your Present Home,* Joe Carter, ed. (Emmaus, Pa.: Rodale Press, 1981).

NOTE: Solar energy data (column B) refers to average conditions, with clear and cloudy days combined.

TABLE 3–2—*continued*			
	Avg. Solar Energy Received on Horizontal Surfaces Annual Total (KBtu/ft²/yr.)		
State or Province	**Ambient Temp. (°F) Annual Avg.**		
	Location	**A**	**B**
Georgia	Atlanta	61.4	555
	Macon	65.6	556
	Rome	60.4	530
	Savannah	66.4	546
Idaho	Boise	51.0	538
	Pocatello	47.0	559
Illinois	Cairo	59.6	529
	Chicago	50.8	468
	Granite City	55.3	500
	Moline	50.1	474
	Springfield	53.6	496
Indiana	Evansville	57.0	515
	Fort Wayne	50.3	480
	Indianapolis	52.1	471
Iowa	Des Moines	49.2	487
	Sioux City	49.1	512
Kansas	Concordia	54.1	533
	Dodge City	55.4	600
	Topeka	54.9	500
	Wichita	57.1	546
Kentucky	Louisville	55.7	484
Louisiana	Lake Charles	68.6	568
	New Orleans	68.6	537
	Shreveport	66.1	554
Maine	Caribou	38.4	432
	Portland	45.0	477
Maryland	Baltimore	55.2	477
Massachusetts	Boston	51.4	423
	Worcester	46.8	434
Michigan	Alpena	42.1	438
	Detroit	50.1	452
	Escanaba	41.9	435
	Grand Rapids	47.6	447
	Lansing	47.6	447
	Marquette	42.6	433
	Sault Ste. Marie	40.6	454
Minnesota	Duluth	37.9	440
	Minneapolis	43.7	445
Mississippi	Jackson	65.5	534
Missouri	Columbia	55.0	520
	Kansas City	56.8	511
	St. Louis	55.3	500
	Springfield	56.5	525

State or Province	**Avg. Solar Energy Received on Horizontal Surfaces Annual Total (KBtu/ft²/yr.)**		
	Ambient Temp. (°F) Annual Avg.		
	Location	**A**	**B**
Montana	Billings	47.5	510
	Glasgow	41.4	518
	Great Falls	44.7	591
	Havre	42.1	560
	Helena	43.4	511
	Missoula	43.2	462
Nebraska	Lincoln	52.8	503
	North Platte	49.2	529
	Omaha	51.5	494
	Valentine	46.9	540
Nevada	Ely	44.3	638
	Las Vegas	65.7	695
	Reno	48.4	645
	Winnemucca	47.4	583
New Hampshire	Concord	45.6	435
New Jersey	Atlantic City	54.1	497
	Trenton	53.9	479
New Mexico	Albuquerque	56.6	707
New York	Albany	47.6	447
	Binghamton	45.8	445
	Buffalo	46.7	447
	New York	54.5	432
	Schenectady	no data	386
North Carolina	Asheville	54.7	542
	Charlotte	60.8	548
	Greenville	no data	597
	Raleigh	59.5	515
	Wilmington	63.8	531
North Dakota	Bismarck	42.2	502
	Devils Lake	40.0	457
	Fargo	41.1	445
	Williston	40.9	467
Ohio	Cincinnati	53.6	479
	Columbus	52.0	438
	Toledo	49.0	468
Oklahoma	Oklahoma City	60.3	595
	Tulsa	59.7	525
Oregon	Eugene	52.5	443
	Medford	52.6	531
	Portland	52.9	391

TABLE 3–2—*continued*		Avg. Solar Energy Received on Horizontal Surfaces Annual Total (KBtu/ft²/yr.)	
State or Province	Location	Ambient Temp. (°F) Annual Avg. A	B
Pennsylvania	Allentown	51.1	461
	Harrisburg	53.3	441
	Philadelphia	53.5	479
	Pittsburgh	53.0	471
Rhode Island	Providence	50.1	462
South Carolina	Charleston	65.2	544
	Columbia	64.0	548
	Spartanburg	61.2	540
South Dakota	Huron	44.7	498
	Rapid City	46.8	537
Tennessee	Chattanooga	61.2	515
	Memphis	61.5	535
	Nashville	60.0	497
	Oak Ridge	58.2	482
Texas, northern	Abilene	64.3	608
	Amarillo	58.7	635
	Dallas	65.8	533
	El Paso	63.3	720
	Fort Worth	65.8	609
Texas, southern	Austin	68.3	554
	Brownsville	73.7	603
	Corpus Christi	71.8	588
	Houston	69.2	561
	San Antonio	68.7	602
Utah	Salt Lake City	50.9	558
Vermont	Burlington	43.2	427
Virginia	Lynchburg	56.7	509
	Norfolk	59.7	515
	Richmond	58.1	501
Washington	Seattle-Tacoma	51.1	394
	Walla Walla	54.2	476
West Virginia	Parkersburg	55.0	454
Wisconsin	Green Bay	44.3	440
	Madison	45.0	460
	Milwaukee	45.1	464
Wyoming	Cheyenne	45.9	581
	Lander	44.4	586
	Sheridan	45.2	522

State or Province	Location	Avg. Solar Energy Received on Horizontal Surfaces Annual Total (KBtu/ft²/yr.) Ambient Temp. (°F) Annual Avg. A	B
Washington, D.C.		57.0	477
Alberta	Calgary	38.2	433
	Edmonton	34.5	381
	Medicine Hat	41.2	456
British Columbia	Prince Rupert	45.8	271
	Vancouver	49.7	387
Manitoba	Brandon	35.1	441
	Winnipeg	36.2	435
	Fredericton	41.8	397
New Brunswick	Moncton	41.6	401
Newfoundland	Gander	39.7	332
	St. Johns	40.8	340
Nova Scotia	Halifax	42.8	393
	Sydney	42.8	387
	Yarmouth	44.6	391
Ontario	London	45.5	431
	Ottawa	42.5	423
	Sudbury	38.5	411
	Toronto	45.5	434
	Timmons	34.6	417
Prince Edward Island	Charlottetown	42.0	402
	Summerside	42.5	410
Quebec	Quebec	39.9	394
	Sherbrooke	39.2	399
Saskatchewan	Prince Albert	32.3	400
	Regina	35.7	439
	Saskatoon	34.9	431

TABLE 3–3

Adjustment Factors For Collector Tilt And Orientation
Ratio of Yearly Total Solar Energy Received on Surfaces of Various Tilt and Orientation Angles to that Received on a Horizontal Surface

Tilt Angle of Collector Surface	Orientation Angle of Collector Surface			
	0° True South	15° East or West of True South	30° East or West of True South	45° East or West of True South
Vertical (90°)	0.72	0.69	0.65	0.52
Latitude +25°	0.98	0.95	0.90	0.73
Latitude +15°	1.03	1.01	0.97	0.78
Latitude	1.10	1.09	1.06	0.86
Latitude −15°	1.10	1.10	1.08	0.88
Horizontal (0°)	1.00	1.00	1.00	1.00

SOURCE: Joe Carter, ed., *Solarizing Your Present Home* (Emmaus, Pa.: Rodale Press, 1981).

NOTE: Factors from these columns are used for worksheet 3–2, line 2.

TABLE 3–4

Glazing Factors for Various Solar Glazing Materials

Glazing Material (generic)	Common Description or Brand Name	Transmittance of a Single Layer (use for worksheet 3–2, lines 5 and 6)
Window glass	double strength	0.84
Low-iron glass	AFG Sunadex Hordis Heliolite	0.91
Fiberglass	Solar Components Sun-Lite Premium II Filon Solar-E Lasco Crystalite-T Glasteel Glasteel	0.85
Acrylic	Du Pont Lucite L or Lucite SAR Rohm & Haas Plexiglas CYRO ACRYLIC GP	0.89
Acrylic (double wall)	CYRO EXOLITE Acrylic	0.83
Polycarbonate	G.E. Lexan Rohm & Haas Tuffak Sheffield Poly-Glaz	0.87
Polycarbonate (double wall)	Rohm & Haas Tuffak-Twinwal Structured Sheets Qualex	0.79
Polyester	Martin LLumar	0.88
Polyester (laminate)	3M Flexigard 7410 or 7415	0.89
Fluorocarbons	Du Pont Teflon FEP Du Pont Tedlar PVF	0.96 0.90

SOURCE: Adapted from *Solarizing Your Present Home*, Joe Carter, ed. (Emmaus, Pa.: Rodale Press, 1981).

NOTE: Glazing Averaging Factor (for worksheet 3–2, line 7) = 0.93 for single-glazed systems; 0.90 for double-glazed systems or double-wall glazings.

TABLE 3–5

Heat Loss Factors
for Batch and Flat Plate Collectors

Description of System		Collector Loss Factor (use for worksheet 3–2, line 10)
Batch collectors	freestanding outside with insulating doors	0.50
	in building envelope with insulating doors	0.50
	in building envelope without insulating doors	0.54
Flat plate collectors	thermosiphon flow	0.52
	pumped flow	0.50

SOURCE: Adapted from *Solarizing Your Present Home*, Joe Carter, ed. (Emmaus, Pa.: Rodale Press, 1981).

NOTES: Glazing Loss Factor (for worksheet 3–2, line 11) = 1.0 for single-glazed collectors; 0.8 for double-glazed collectors or double-wall glazings. Absorber Loss Factor (for worksheet 3–2, line 12) = 1.0 for nonselective absorbers; 0.9 for selective absorbers (black chrome, etc.). Heat Exchanger Loss Factor (for worksheet 3–2, line 13) = 1.0 for all systems without heat exchangers; 1.1 for systems with heat exchangers.

TABLE 3–6

Diminishing Returns Factors

If your gross solar fraction (from worksheet 3–2, line 17) is:	then for DHW systems your net solar fraction (enter on worksheet 3–2, line 18) is:
0.05	0.05
0.10	0.10
0.15	0.15
0.20	0.20
0.25	0.25
0.30	0.30
0.35	0.35
0.40	0.40
0.45	0.45
0.50	0.49
0.55	0.53
0.60	0.56
0.65	0.59
0.70	0.61
0.75	0.64
0.80	0.66
0.85	0.68
0.90	0.69
0.95	0.71
1.00	0.72
1.10	0.74
1.20	0.76
1.30	0.78
1.40	0.80
1.50	0.82
1.60	0.84
1.70	0.86
1.80	0.88
1.90	0.90
2.00	0.92

SOURCE: Joe Carter, ed., *Solarizing Your Present Home* (Emmaus, Pa.: Rodale Press, 1981).

TABLE 3–7

Finding Your Energy Cost per Million Btu's

Listed below are some conversion factors for fuels that are commonly used to heat water. We've thrown in wood and coal for you multi-fuel users. All you have to do to finish the equation is plug in the price you're paying per unit of the fuel you use to heat your water. The most accurate way to determine your unit cost is simply to look at the total amount of your bill and divide it by the number of units you're being billed for. Then multiply unit cost times the energy cost factor to get the cost per MBtu.

Fuel	Cost ($) per Fuel Unit (Btu per unit; assumed system efficiency, %)	× Energy Cost Factor	= Cost ($ per MBtu)	Fuel	Cost ($) per Fuel Unit (Btu per unit; assumed system efficiency, %)	× Energy Cost Factor	= Cost ($ per MBtu)
Fuel oil	$/gal. (139,600 Btu/ gal.; 65%)	11.03	_____	Wood	$/cord (average 20,000,000 Btu/cord seasoned softwood-hardwood mix; 50%)	0.10	_____
Electricity	$/KWH (3413 Btu/ KWH; 100%)	293.00	_____				
Natural gas	$/therm (100,000 Btu/ therm; 70%)	14.29	_____	Coal	$/ton (30,000,000 Btu/ton; 50%)	0.07	_____
Propane	$/gal. (94,000 Btu/ gal.; 70%)	15.20	_____				

SOURCE: Joe Carter, ed., *Solarizing Your Present Home* (Emmaus, Pa.: Rodale Press, 1981).

4

BUILDING, INSTALLING AND USING BATCH COLLECTORS

As it was described previously, a batch collector consists of a water tank that has been encased in a glazed, insulated box that is placed in the sunshine to catch solar energy. Water is heated and stored in the tank, and the hot water is drawn away on demand.

This collector and the plumbing system that connects it to your water heater are both essentially simple to put together and use. It's a passive system, which means it needs no pump or electrical controls. But for some this simplicity might be the batch collector's biggest drawback. Unlike more complicated, and expensive, systems, batch collectors usually have little or no automatic freeze protection. In very cold climates, a batch collector must either be closed down for the winter or its owner must perform the daily chore of opening and closing insulated doors so that the water tank and the exposed plumbing won't freeze. But batch systems are nonetheless attractive to many because they work, because they're self-operating and because of their low cost. A good batch heater can be built for about $500.

Although batch collectors have some basic design requirements, they can be made in many different shapes and sizes to suit specific applications. Some large units encase as many as four water tanks. They can be installed on the lawn, up next to the house, into a roof opening, even inside a solar greenhouse.

The bulk of this chapter explains how you can build a basic, one-tank batch collector. Later in this chapter you'll also see some design variations, such as collectors with multiple water tanks, or a collector that can be built into a roof. Even these variations aren't the only ones that are possible. The batch heater is basically just a type of system that can be built in just about any way that fits with your house. The phrase "batch collector" defines an idea, rather than a rigid design. The idea is to place a water storage tank in the sun, behind some glazing, where it can directly absorb solar energy. The box that encloses the water tank should admit as much sunlight as possible to the surface of the tank, and the box should also be well insulated to retain the valuable heat. Many different kinds and sizes of water tanks have been used. Feel free to experiment and build the batch collector that's best for your home.

Building a Cusp Reflector Batch Heater

The collector described here is a *cusp reflector* batch collector. This is a tried-and-true solar water heater that was developed and studied by Rodale's Research and Development Department. The curved cusp reflector is a mirrorlike device that lines the collector box behind the water tank. The cusp reflects sunlight to the entire surface of the water tank, all day long (figure 4–1). Without the reflector most of the tank's surface would be shaded and less water would be heated.

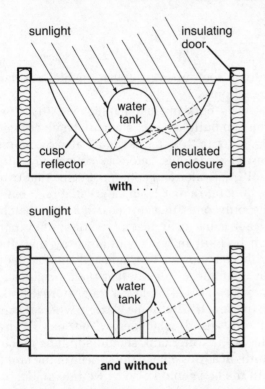

with . . .

and without

Figure 4–1: Compared to a box that is simply lined with reflective material, the geometry of a cusp reflector causes 10 to 15% more light to be bounced onto the batch tank.

This design also includes ways to reduce the collector's vulnerability to freezing. Insulated hatch covers can also serve as exterior reflectors to bounce more sunlight through the glazing. Exposed plumbing is protected with heavy insulation and thermostatically controlled "pipe tape" or "heat tape," which

heats up when an electric current is passed through it. Another freeze protection option depends on how and where the batch collector is installed. If it's installed on the side or on the roof of a house, there can be a vent connection between the house and the collector. On very cold nights the vent is opened to borrow a little house heat to keep the tank and pipes from freezing. All these features can make batch systems appropriate for all but the most frigid climates.

The following construction steps describe a basic, boxlike collector that can be installed just about anywhere. It should be stressed again, though, that "batch" is really a system and the collector enclosure and style of installation are elements that can be customized to your specific application.

Making the Cusp Reflector

The first step here is to make the frame that will hold the cusp reflector. The frame is made from pieces of ½-inch AC-grade exterior plywood and ⅛-inch hardboard (see the materials list for complete materials and

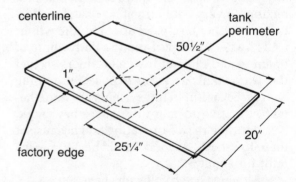

Figure 4–2: Two reflector end frames are made from 20 x 51-inch pieces of plywood. A centerline is drawn down the middle of one piece. Then a mark is made 1 inch up from the factory edge. This is where the water tank is placed for the tracing of its perimeter.

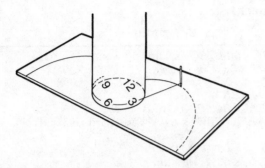

Figure 4–3: Here we see the cusp being drawn, one side at a time. A pencil tied to a wire is set at the 12 o'clock point. The wire is drawn tightly around the perimeter of the tank and its other end is taped to the 9 o'clock point. When the pencil is swung away, the curve is automatically drawn. The process is reversed to draw the other curve.

dimensions). Cut two pieces of plywood to 20 by 50½ inches. These will be used for the ends of the reflector frame and will be cut to the curves needed to form the cusp.

Set one of the plywood pieces on a smooth, level surface. The smooth side of the plywood should be up, and the factory edge should face toward you. Referring to figure 4–2, find the center of the 50½-inch side and draw a line through the center, perpendicular to the long edge. On this centerline, 1 inch above the factory edge, make a mark with a pencil. This mark will be used in the next step, in which you have to work with the batch tank.

It's recommended that you buy a 40-gallon range boiler tank with extra-heavy galvanizing for your batch tank. It will probably cost $100 to $120, and will last about as long as the average water heater, about 10 years. To find one, check with a major plumbing retailer or wholesaler in your area. Stainless steel tanks that are custom-made for batch collectors are available for about $250 to $400,

and should last much longer, perhaps up to 40 years. Copper can be even more expensive. Buy the best tank that you can afford. You could use a glass-lined steel tank of the type used in water heaters, but it might not work as well because the glass lining adds extra resistance to heat gain. Used water heater tanks are available at a low price, but you can never tell how long they'll last before springing a leak.

Once you have acquired the water tank (a list of suppliers can be found in Appendix 1), measure its diameter. Most 40-gallon range boilers have a diameter of about 14 inches. After measuring the diameter, lightly draw two lines parallel with the centerline on the piece of plywood, with each line being half the tank's diameter away from the centerline (figure 4–2).

Now stand the tank on the plywood, placing the edge of the tank on the mark that is 1 inch from the factory edge and on the two parallel lines that mark the tank's diameter.

Now imagine, if you will, that the tank's circumference is divided like the face of a clock, and that the centerline runs beneath the tank at 12 and 6 o'clock. Find a flexible metal wire about 4 feet long, tie a knot in one end and slip the knot around a pencil. Place the pencil point next to the tank on the centerline at the 12 o'clock position. Wrap the wire clockwise around the tank and firmly tape the other end at approximately the 9 o'clock position. With the wire taut, swing the pencil away from the tank off the factory edge, as shown in figure 4–3, and you'll draw the proper curve for the reflector. (It's important that you use a wire for this step, as string can stretch and make an inaccurate curve.) Now remove the wire and reattach it near the 3 o'clock position. Repeat the pro-

Materials Checklist

Item	Quantity	Description
Involute pattern blanks	2	½" x 20" x 50½" AC exterior plywood
Involute top cleats	2	¾" x 3½" x 53⅞" #2 pine
Involute center cleats	2	½" x 4" x 9" AC exterior plywood
Tank support bracket cleats	4	¾" x 3½" x 15" #2 pine
Reflector side boards	2	1½" x 3½" x 84⁹⁄₁₆" KD studs
Reflector hardboard backing sheets	2	⅛" x "AA" or "BB" x 84½" hardboard (see text)
Reflector center support	1	1½" x 1½" x 84⁹⁄₁₆" KD studs
End frame studs	4	1½" x 3½" x 53⅞" KD studs
End frame studs	8	1½" x 3½" x 19½" KD studs
Side and back frame studs	3	1½" x 3½" x 86¹⁄₁₆" KD studs
Side frame studs	2	1½" x 3½" x 15½" KD studs
Insulation	1 roll	3½" x 23" x 70' paper-backed fiberglass insulation
Reflector back panels	2	¼" x 26¹⁵⁄₁₆" x 93¹⁄₁₆" AC exterior plywood
Glazing frame pieces	6	¾" x 1⅝" x 86⅛" #2 pine
Glazing frame pieces	6	¾" x 1⅝" x 53⅜" #2 pine
High-quality paint primer	1 qt.	For oil-based paints
Paint	1 qt.	Oil-based porch and deck enamel (any color)
Aluminum flashing	2	.010" x 4" x 83½" aluminum flashing
Nails	1 lb.	1" underlayment nails
Tank supports	2	³⁄₃₂" x 1½" x 1½" x 83¼" slotted angle iron
Tank support cross-members	2	³⁄₃₂" x 1½" x 1½" x 11⅛" slotted angle iron
Washers	26	⅜" flat washers
Tank support brackets	4	³⁄₃₂" x 1½" x 1½" x 1½" slotted angle iron
Nuts	12	#16 x ⅜" hex nuts
Bolts	11	⅜"–16 x ¾" hex head
Lag bolts	4	⅜" x 3" hex head
Wood glue	1	4 oz. carpenter's wood glue
Metal wire	1 roll	Thin steel wire
Nails	2 lb.	12d common nails
Nails	1 lb.	8d common nails
Nails	6 oz.	1" paneling nails
Sealer	2 qt.	Exterior polyurethane varnish
Joist hangers	4	2" x 4" joist hangers
Truss plates	10	1½" x 6" truss plates
Pipe strap	5'	¹⁄₁₆" x 1⅜" pipe strap
Screws	36	#6 x ¾" flathead wood screws
Screws	20	#8 x 2½" flathead wood screws
Screws	24	#8 x 1½" brass flathead wood screws
Staples	1 pkg.	¼" staples
Adhesive	2 tubes	Trim adhesive
Selective coating foil for tank	11½ L/F	2 mil x 24" x 11½' (pressure-sensitive adhesive type)
Inner glazing layers (film)	42 L/F	2 pieces of 1 mil x 50" x 21' Teflon glazing film
Outer glazing layer (glass)	30'²	2 pieces of ³⁄₁₆" x 51½" x 42" glass
Butyl glazing tape	64'	⅜" wide preshimmed glazing tape
Solar reflective Mylar	25 L/F	5 mil x 48" x 25'

Materials Checklist—*continued*

Item	Quantity	Description
Aluminum foil duct tape	1 roll	2 mil x 2½" x 180'
Clear silicone caulk	2 tubes	High-quality silicone caulk
Hard copper tubing	77½"	½" x 77½"
Copper wire	2"	Copper wire
Copper Cap	2	½" cap
Reducing bushing	2	1" MPT x ½" FPT threaded bushing (brass)
Reducing bushing	3	1" MPT x ¾" FPT threaded bushing (brass)
Compression fitting	2	¾" MPT x ⅝" compression fitting
Copper unions	2	½" x ½" unions
Reducing bushing	1	¾" MPT x ½" FPT bushing (galvanized)
Bend	1	½" MPT x ½" copper elbow
Vacuum valve	1	½" Watts 36 A
P & t valve	1	¾" pressure-and-temperature valve
Pipe plug	1	1" galvanized pipe plug
Hot tube jacket	1	¾" i.d. x 57" CPVC tubing
Water tank	1	1 40-gal. range boiler tank, extra-heavy galvanized
Tubular insulation	12'	⅝" i.d. x ¾" wall
Heat tape	30'	Heating tape
PVC pipe	36"	4" diameter PVC solid drainpipe
P & t valve drain tube	1	½" x 24" CPVC tubing
Nipple	1	½" MPT x ½" MPT
Elbow	1	½" FPT x ½" FPT, 45° bend
CPVC adapter	1	½" MPT x ½" hub
Plumber's strap	8 pcs.	Plumber's strap
Denatured alcohol	1 qt.	Denatured alcohol (for cleaning tank)
Black graphite paint	1 qt.	Sun In paint
Teflon tape	1 roll	Thread-wrapping tape

SOURCE: Adapted from "The Best We Know" by Frederic S. Langa (*New Shelter*, July/August, 1981).

cedure to draw an identical curve on the other side of the plywood.

Trace the tank's circumference and remove the tank from the plywood. Now you have the basic outline of the reflector curve inscribed on the plywood. The next few steps will "fine-tune" the shape of the curve to improve the reflector's performance and make it easier to work with.

Referring to figure 4–4, use a straightedge or a yardstick to draw a line from each corner of the plywood factory edge to the point of tangency with the curve. This wid-

ens the reflector and will eventually permit more light to strike the water tank.

Position the outer corner of a carpenter's framing square between the humps at the top of the curve (figure 4–4), and mark the 90-degree angle with a pencil. This angle will become the new cut line where the two curves come together. The original peak is altered in this way so that the reflector can be more easily installed inside the collector box.

Next, cut four 1½-inch-long pieces of the slotted angle iron. These will be used as support brackets for the water tank. Referring to

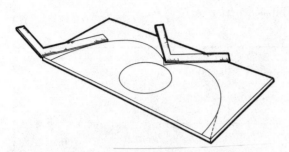

Figure 4–4: For proper construction the cusp must be altered in three places. Use a framing square to change the valley at the top to a 90° angle. Then use a straightedge to alter the two bottom edges of the curves so that they extend to the corners of the plywood. This widens the reflector cavity to admit more sunlight.

figure 4–5, fasten together two angle iron pieces using a ⅜–16 x ¾-inch-long hex head bolt and nut with two ⅜-inch flat washers. As shown in the figure, part "A" of the angle iron assembly is the tank support bracket while part "B" simulates a longer tank support that will be added later.

Next measure and mark along the plywood factory edge 5 inches from both sides of the centerline (figure 4–6). These lines will help in positioning the fastening holes for the tank support brackets.

Place part "A" of the angle iron assembly on one of the 5-inch lines. Position part "B" so that both "legs" of the angle iron touch the tank circle. Mark the fastening hole location, as shown in figure 4–6. Repeat this procedure on the other side of the tank circle. As the figure indicates, you'll have to reverse part "B" of the tank support bracket to do this. You should now have the complete pattern for one end of the reflector frame.

Set the marked plywood on top of the second sheet of plywood. The rough sides should face each other and both factory edges

should be toward you. Fasten the sheets together with C-clamps. With a saber saw, cut both sheets of plywood along the curved reflector lines. Save the plywood drop-offs; they'll be used in a later step. Next drill the ⅜-inch-diameter tank support bracket holes. You should now have two identical reflector frame ends. Mark the smooth side of one of the ends "A" and the other one "B", and set them aside.

Unroll the aluminized Mylar onto a smooth, flat surface that has been covered with newspaper. (The newspaper will help prevent scratches.) The plastic side of the Mylar should face the newspaper; you can tell which side of the Mylar is plastic and which is aluminum by rubbing a pencil eraser on one side. If the eraser turns black, it's been rubbed across the aluminum side.

With the Mylar properly rolled out, put the "A" piece of plywood, smooth side up, on it. Kneel on the plywood pattern and, with a utility knife, cut the Mylar around the edges of the pattern. Mark an "A" on the aluminum

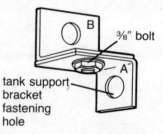

Figure 4–5: The tank support brackets are made as shown with angle iron and ⅜" bolts. Part "A" will later be bolted to one of the reflector end frames. Part "B" merely simulates a tank support and will be removed later to make room for the real thing.

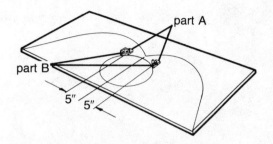

Figure 4–6: Two tank support brackets are placed on each reflector end frame as shown. The tank support bracket holes will be drilled 5 inches from each side of the centerline.

side of the Mylar (the side that's facing up). Cut out another piece of Mylar using the "B" piece of plywood, and mark a "B" on the Mylar. Roll up the two pieces of Mylar and set them aside.

The reflector frame ends are the "keystones" of the batch collector in that they form the reflector curvature for the length of the collector. The next steps involve attaching top cleats, center cleats and tank support bracket cleats to the reflector frame ends. These cleats will help connect the reflector frame ends with the rest of the collector.

Attach the top cleats first. Cut two 1 x 4's of #2 pine to a length of 53⅞ inches. Referring to figure 4–7, fasten a top cleat to the rough side of each reflector end frame, flush with the straight edge of the plywood. The cleat will extend beyond each side of the end frame, and this overhang should be equal on both sides. Fasten the cleat first with wood glue and then with 1-inch-long underlayment nails. Drive the nails in from the plywood side.

Fabricate two center cleats according to the dimensions found in the materials list. These can be made from the scrap plywood left over from the reflector end frames. The

center cleats will be used to fasten a 1½ by 1½-inch (2 x 2 nominal) center support beam beneath the cusp reflector, so a 1½ by 1½-inch square should be made by extending the triangular tip of the center cleat 1½ inches beyond the 90-degree valley of the reflector frame end (figure 4–7). Like the top cleats, a center cleat should be fixed to the rough side of each plywood end frame. Use wood glue and 1-inch underlayment nails.

Cut four 1 x 4's to 15-inch lengths. These are the tank support bracket cleats, and they will eventually carry much of the weight of the water tank. Attach the tank support bracket cleats to the rough side of the plywood end frame. As you can see from figure 4–7, the tops of the tank support bracket cleats should be flush against the top cleat. They should also be positioned so that they squarely cover the ⅜-inch tank support bracket holes. Like the other cleats, the tank support bracket cleats should be fastened with wood glue and 1-inch underlayment nails.

Now drill ⅜-inch-diameter holes through the tank support bracket cleats, using as guides the original ⅜-inch tank support bracket holes. Drill a total of four holes, one through each tank support bracket cleat.

With a piece of flexible wire, measure the length of the right or left half of the cusp curve on one of the plywood end frames (figure 4–8). Subtract ⅛ inch from this measurement and call it "AA." Next measure the other half of the same plywood end frame. Call this measurement "BB," but do not subtract ⅛ inch from the "BB" measurement. These measurements will be used in the following step.

Cut a standard sheet of ⅛-inch hardboard "AA" by 84½ inches. The smooth side of this piece of hardboard should be lettered "AA." Cut another sheet of ⅛-inch hard-

Photos 4–1, 4–2: Reflective Mylar pieces are cut with a utility knife, top, *then set in place before bonding to ensure a proper fit,* bottom.

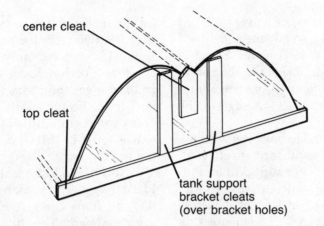

Figure 4–7: The top cleat, center cleat and the tank support bracket cleats are nailed securely to the plywood end frames. The tank support bracket cleats are nailed directly over the bracket holes. Note that the triangular point of the center cleat and the 90° valley of the cusp make a square, to which the reflector center support will be nailed. The center cleats should be at least 5 inches wide and 10 inches long; the triangular, 90° point can be cut with a table saw. Also, notice that the top cleat extends beyond both sides of the plywood. Later, the reflector side boards will be nailed to the top cleats.

board "BB" by 84½ inches. This piece of hardboard should be lettered "BB," also on its smooth side.

On a flat surface that's been covered with newspaper, unroll a section of reflective Mylar, with the plastic side facing down. Cover the Mylar with the hardboard piece lettered "AA." The smooth side should be facing up. With a utility knife, cut the Mylar 4 inches larger in width than the hardboard sheet, but 1 1/16 inches shorter than the length of the hardboard. This will make the length of this section of Mylar 83 7/16 inches, while the width will be "AA" plus 4 inches. On the aluminum side, mark this piece of Mylar "AA." Cut another piece of Mylar with the hardboard sheet lettered "BB," using the same procedure, and letter this piece "BB." Roll up these pieces, use tape or a rubber band to prevent them from unrolling, and set them aside.

Now you're ready to make the reflector center support. Cut one 2 x 2 to a length of 84 9/16 inches. Lay this piece on a flat surface, and set hardboard sheet "AA" on top of it, with the smooth side of the hardboard facing up. Note that the center support is 1/16 inch longer than the hardboard sheet. Position the hardboard on the center support so that the

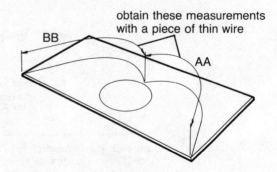

Figure 4–8: The length of each cusp curve is measured with a piece of wire.

center support extends ¹⁄₃₂ inch over each end of the hardboard. Using wood glue and 1-inch underlayment nails spaced about 2 inches apart, nail the hardboard sheet to the center support, referring to figure 4–9. The center support should be flush to the edge of the hardboard.

You'll need a friend to help you with the next step. Turn over hardboard sheet "AA" so that the reflector center support faces up (figure 4–9). Position hardboard sheet "BB" against the center support so that it is perpendicular to sheet "AA." The smooth side of hardboard sheet "BB" should be facing away from sheet "AA." Your helper will have to hold hardboard sheet "BB" while you center it and also stand on sheet "AA" when you glue and nail sheet "BB" to the reflector center support. These hardboard sheets and the center support will hereafter be referred to as the reflector backing assembly.

Cut two 2 x 4 kiln-dried wood sheets to a length of 84⁹⁄₁₆ inches. These will be used for the reflector side boards, which will connect the two reflector end frames to the reflector backing assembly.

Stand both reflector end frames on their factory edges, and attach the reflector side boards to the top cleats of the end frames, using wood glue and 8d common nails. Two nails should be driven per joint. Nail in from the top cleats.

Now the collector is starting to take shape. Place the reflector backing assembly between the end frames. Be careful *not* to install it upside down: The two hardboard sheets should point upward like butterfly wings, not downward like the walls of a tent (figure 4–10). Maneuver the center support of the backing assembly between the two center cleats, as shown in the figure. One person should hold the hardboard sheets of the reflector backing assembly while the other nails the center support to the center cleats. Drive two 8d common nails into each joint.

Allow the reflector backing sheets to

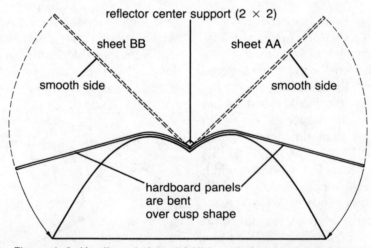

Figure 4–9: Hardboard sheet "AA" is nailed to the reflector center support first, smooth side facing out. Then sheet "AA" is laid flat, smooth side down, and hardboard sheet "BB" is nailed in place, smooth side out.

droop into place around the curves of the end frames, making sure the sheets fit properly. Slowly press the sheets down along the curved edges of the end frames and tuck them inside the side boards to check the final fit (figure 4–9).

Now lift the hardboard sheets back up. Working on one sheet at a time, apply wood glue along the plywood edge on both end frames, and let the sheet droop back down. Guide the hardboard along the ½-inch curved edge, and tuck it inside the side board. Nail the ends to the edges of the plywood end frames with 1-inch-long paneling nails. Begin nailing at the center support, fastening both ends of the sheet at the same time. Then follow the same procedure to glue and nail the other hardboard sheet into place. The reflector backing assembly has now been securely nailed to the collector frame.

Seal the rough side of the hardboard and the plywood end frames with two coats of exterior-grade polyurethane. Don't coat the smooth sides of the hardboard or the plywood, as Mylar will cover these surfaces. Make sure the polyurethane dries completely before you proceed.

While you're waiting for the polyurethane to dry, you can cut four 2 x 4's to a length of 53⅞ inches. These are the long end frame studs (figure 4–11). Then cut eight 2 x 4's at 19½ inches. These are the short end frame studs. With these pieces put together the two end frame assemblies. On each end frame assembly, two of the shorter studs should be spaced so that the tank support bracket cleats on the reflector end frames can be fastened to them. These four short studs (two on each end frame assembly) are nailed to the end frame assemblies with joist hangers, as well as 12d common nails, which are

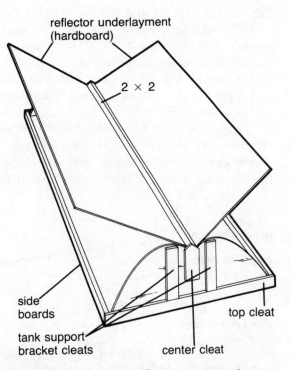

reflector underlayment (hardboard)

2 × 2

side boards

top cleat

tank support bracket cleats

center cleat

Figure 4–10: Now the collector starts to take shape. The side boards are nailed to the top cleat (drive the nails through the top cleats into the side boards), then a 2 x 2 reflector center support is nailed to the center cleats. The hardboards should point upward like the wings of a butterfly, and not downward like a tent.

driven through the longer studs. Joist hangers are the metal brackets shown in figure 4–11. The other studs comprising the end frame assembly can be nailed in place with 12d common nails, two in each joint.

Once the end frame assemblies have been built, lay one of them down on a flat surface. Tilt it up so it's vertical and set it on the end frame. With a framing square, line up the straight edge of the plywood end frame with the end frame studs, and when everything's right, fasten the reflector assembly to the end frame studs with 8d common nails. Drive the nails through the ½-inch plywood, through the top cleat, into the end frame stud. Also drive nails through the plywood, through the tank support bracket cleats and into the short end frame studs that have been reinforced with joist hangers.

Invert the collector and fasten the other end frame stud assembly in the same way. As you can see, what's beginning to take shape is a rugged, yet simply constructed, box that will be able to hold the weight of 500 pounds of filled water tank.

Next cut three 2 x 4's at 86¹⁄₁₆ inches. Two of these studs will be reflector side boards. Fasten them to the bottoms of the end frame stud assemblies, as shown in photo 4–3. The third stud should be fastened in the middle of the back. Nail these three studs in place with 8d common nails, 1-inch underlayment nails and truss plates, as shown in figure 4–11. (A truss plate is a reinforcing

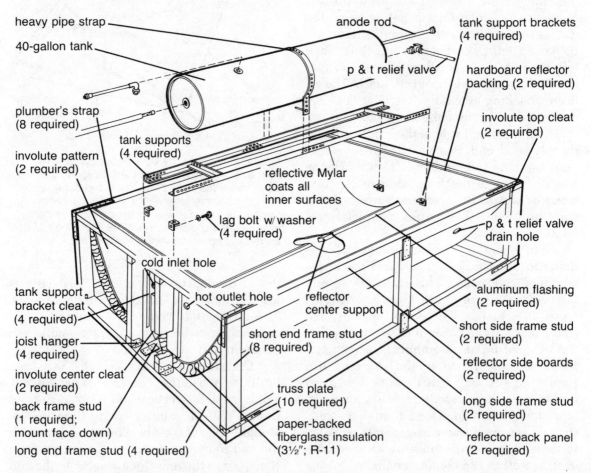

Figure 4–11: This illustration shows all parts of the collector.

Photo 4–3: Turned on its side, the side boards and long side frame studs have already been nailed to the end frame stud assemblies. Now, using truss plates, the short side frame studs are installed.

piece of drilled metal through which nails are driven.) Plumber's strap should now be used to reinforce all the corners of the end frame studs.

Cut two 2 x 4's at 15½ inches. These are the short side frame studs, which are installed in the middle of the sides of the collector, using truss plates at the top and bottom (photo 4–3).

Now you can build the tank supports. These will be made from the 1½-inch angle iron. Cut two 83¼-inch pieces and two 11⅛-inch pieces for the cross-members. Assemble the tank supports with ⅜–16 x ¾-inch hex head bolts, washers and nuts. The cross-members should be spaced at a distance from each other corresponding to the length of your 40-gallon water tank, dividing that length into thirds (figure 4–11).

Referring again to figure 4–11, attach a 5-foot length of the pipe strap to the middle of the tank supports. Shape the strap to the contour of the water tank before bolting the strap to the angle iron. Use ⅜–16 x ¾-inch hex head bolts, washers and nuts to fasten the strap to both of the 83¼-inch-long angle iron supports. For a moment, set the tank supports aside.

Now it's time to prepare the water tank for painting. Examine its surface and sand down any rough spots. To remove grease or dirt, wipe the tank with solvents such as alcohol, lacquer thinner and mineral spirits.

The next step is to cover the tank with a selective coating. Regular flat black paint can be used, but it won't absorb solar energy or retain heat as well as a selective coating. Selective coating tapes, or foils, with brand

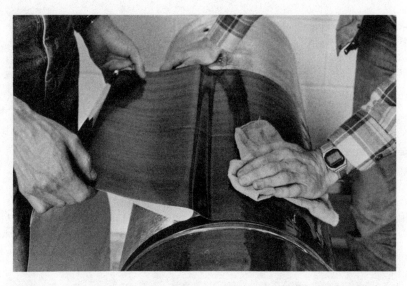

Photo 4–4: A soft cloth is used to smooth out the adhesive-backed selective coating foil.

names like Sunsponge and Maxorb, have a special black surface for maximum absorption and minimum reradiation of heat from the tank. This is an important consideration with batch collectors, since the stored hot water is always sitting outside. An alternative to the selective surface foils is a paint, known as thermalox, that has similar qualities.

You'll need a helper when applying the selective coating self-adhesive foil. First lay the tank on sawhorses. Measure its circumference and add an inch to that measurement. Slice off a corresponding length of the foil with a utility knife, using a framing square as a cutting guide.

Expose a 3- to 4-inch strip of the foil's adhesive backing. Using the tank's bottom edge as a guide, and starting at the welded seam, carefully apply the foil (photo 4–4). Your helper can peel away the adhesive backing paper while you smooth out the foil with a soft rag. When you come to one of the tank's plumbing ports, the foil can be neatly

cut away with a utility knife to expose the port. When you've wrapped the foil all the way around the tank, peel away a short strip of the protective plastic sheet (which protects the exposed selective coating until you install the tank in the collector box) and stick the adhesive side directly onto the black selective surface. (This is why you cut the foil longer than the tank's circumference.) Examine your work with a critical eye. A wallpaper seam roller or a soft rag can be used to smooth down the foil and make a total bond with the tank. Apply another sheet of foil, again starting at the welded seam. The second sheet of foil should be overlapped ½ to 1 inch onto the first sheet. This sheet is applied and smoothed out in the same way the first sheet was applied. Continue covering the tank until it's all covered.

Coat both ends of the tank with black graphite paint. Follow the manufacturer's instructions when applying the paint, taking care *not* to paint the pipe threads in the plumbing ports. While you've got the paint out, coat the tank supports and the 1½-inch tank support brackets.

After the paint has dried, place the tank supports on the sawhorses. Maneuver the tank's welded seam between the two sections of slotted angle iron. The tank's concave bottom should fit snugly against the small, 11⅛-inch piece of angle iron at the bottom of the support assembly.

Pull the heavy pipe strap around the tank, shaping it so it neatly encircles the tank. Bolt the free end of the strap to the tank supports. If the strap is too long, cut off the excess.

Plumbing the Tank

Assemble the hot water outlet tube and the cold water inlet tube according to the instructions in figure 4–12, and install the proper plumbing fittings into the tank ports

as shown. The threads should be sealed with Teflon plumbing tape, you may not be familiar with the *anode rod* that is screwed into one of the ports in the top of the tank. The rod itself is a magnesium alloy, and it serves as a "sacrificial" device that staves off excessive corrosion of the tank. The rod itself corrodes instead. All water heaters have them, but they aren't necessarily that easy to find. You'll have to call around to various plumbing suppliers, and a special order may be required, but at about $10 for the rod the investment in longer tank life is well worth the expense.

Install the four small tank support brackets. (These are the 1½-inch pieces of angle iron that you cut many steps ago.) Bolt them through the ⅜-inch tank support bracket holes (figure 4–11) with ⅜ x 3-inch lag bolts and ⅜-inch flat washers.

Photo 4–5: Angle iron sections are bolted together to support the tank. A length of plumber's strap is drawn around the tank's midsection, then bolted to the angle iron to secure the tank. Then the tank is plumbed.

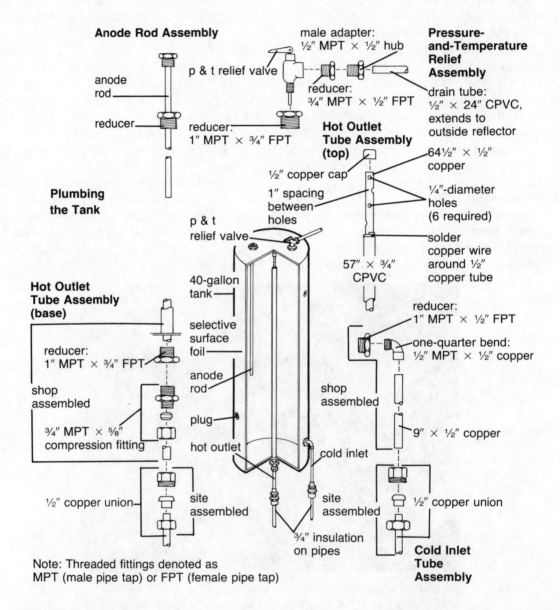

Anode Rod Assembly

anode rod

reducer

p & t relief valve

reducer:
1″ MPT × ¾″ FPT

Plumbing the Tank

male adapter:
½″ MPT × ½″ hub

reducer:
¾″ MPT × ½″ FPT

Pressure-and-Temperature Relief Assembly

drain tube:
½″ × 24″ CPVC,
extends to
outside reflector

Hot Outlet Tube Assembly (top)

½″ copper cap

1″ spacing between holes

64½″ × ½″ copper

¼″-diameter holes
(6 required)

solder copper wire around ½″ copper tube

p & t relief valve

57″ × ¾″ CPVC

Hot Outlet Tube Assembly (base)

reducer:
1″ MPT × ¾″ FPT

shop assembled

¾″ MPT × ⅝″ compression fitting

40-gallon tank

selective surface foil

anode rod

plug

hot outlet

reducer:
1″ MPT × ½″ FPT

one-quarter bend:
½″ MPT × ½″ copper

shop assembled

9″ × ½″ copper

½″ copper union

site assembled

cold inlet

site assembled

½″ copper union

¾″ insulation on pipes

Cold Inlet Tube Assembly

Note: Threaded fittings denoted as
MPT (male pipe tap) or FPT (female pipe tap)

Figure 4–12: The tank is plumbed as shown. The hot outlet tube assembly serves to maintain good temperature stratification.

Photo 4–6: The tank and its supports are temporarily set in place in the reflector cavity. In an earlier step the peak of the cusp was altered to make installation easier. Now, with the tank in place, aluminum flashing is used to extend the peak of the cusp right up to the underside of the tank. Then the tank and the supports are removed and the reflective Mylar is installed.

Put the tank supports and the tank temporarily into the reflector cavity. The tank supports should be set on top of the tank support brackets and temporarily fastened down with four ⅜–16 x ¾-inch head bolts, ⅜–16 hex nuts and flat washers.

With everything in place you'll be able to see where you should drill the holes for the cold water inlet and hot water outlet pipes and the drain tube from the pressure-and-temperature relief valve. If you have trouble visualizing where these holes should penetrate the reflector cavity, consult figure 4–11.

Once you have located and marked where the holes should be, remove the tank and the tank supports, and drill the holes. The size of the holes depends on the diameter of the unions that must fit through them, so be sure to measure them to find the right hole size.

After the holes have been drilled, reposition the tank and the tank supports onto the tank support brackets. You'll notice that the peak of the cusp does not touch the bottom of the tank. This is because, in an earlier step, the peak was altered. Now, using aluminum flashing, you can return the peak to its original cusp shape. Cut two pieces of 4 x 83½-inch aluminum flashing. Give the flashing a slight concavity lengthwise by laying a pipe or a rolling pin on a flat surface and rolling the flashing gently over it. Wear gloves while doing this to avoid little cuts.

Using the underside of the water tank as a guide, slide one of the flashing pieces along hardboard sheet "AA" so that it completely closes the gap between the underside of the tank and the hardboard (photo 4–6). Fasten the flashing to the reflector center support with a staple gun. Position and fasten the other piece of flashing to hardboard sheet "BB" the same way. Remove the water tank and the tank support, and tape the two pieces of flashing together at the peak with the aluminum tape.

Now you can cover the reflector cavity with the pieces of Mylar that you cut earlier. The hardboard and plywood surfaces of the reflector cavity should be clean, smooth and dust free. It's also a good idea not to work in direct sunlight.

Install the large sheets first. Start by laying sheet "AA" on hardboard sheet "AA." Make sure the aluminized side of the Mylar is facing up. Flatten out the Mylar, and trim it lengthwise, so that it fits between the plywood end frames, but don't trim the width yet. Position the trim Mylar sheet "BB" the same way.

Don't try to fasten a whole sheet of Mylar at once; work in sections. You'll probably need a helper. Roll Mylar sheet "AA" away from the aluminum flashing peak and, working from the outer edge of the reflector frame, apply a uniform coat of spray adhesive (a spray-on autotrim adhesive is recommended, such as 3M type 08080) along a 12- to 16-inch-wide length of the outer edge of hardboard sheet "AA." Slowly roll the Mylar down onto the adhesive, smoothing out bubbles with a clean, short-nap paint roller. If you come across a bubble that won't go away, you can slit the bubble and smooth it with the rag. Bond the entire sheet of Mylar with the adhesive and don't trim the excess off the peak of the cusp until the adhesive has fully dried. Once the Mylar sheet has been fastened, cover it with newspaper to protect it from the adhesive spray. Mylar sheet "BB" is installed the same way.

When both large Mylar sheets have been glued down, turn the collector onto one of its ends. You can apply the Mylar onto the plywood ends in one step. First, remove the tank support brackets and place the proper sheet of Mylar onto the plywood. Make sure it fits and then remove the Mylar. Coat the entire plywood end with spray adhesive and immediately install the Mylar, smoothing it out with the rag. Turn the collector onto its other end and glue in the last piece of Mylar. Remove the newspaper.

After the adhesive has dried, trim off any excess Mylar from the peak of the cusp and the edges, and cut away the Mylar that's covering the plumbing holes and the tank support bracket fastening holes. Caulk every seam inside the reflector cavity with high-quality silicone caulk.

Before the plywood sheathing goes on the collector box, insulation must be in-

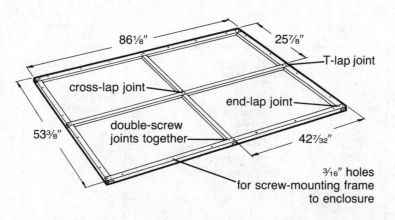

Figure 4–13: All the wooden parts of the glazing and spacer frames are made from 1 x 2's. All the corner, tee and cross joints are lapped, glued and double-screwed.

stalled. You'll need a 3½ by 23-inch by 70-foot roll of paper-backed fiberglass insulation, which will provide an insulation value of R-11. Stuff the insulation all around the back surface of the hardboard and plywood with the paper side facing out (figure 4–11). Staple the insulation to the wood studs. Don't insulate the end of the collector with the three plumbing holes. This end will be insulated after the plumbing has been installed.

The back of the collector is sheathed with ¼-inch plywood (AC exterior grade). Two pieces cut to 26¹⁵⁄₁₆ x 93¹⁄₁₆ inches are needed. They're glued and nailed with 1-inch underlayment nails.

The batch collector's glazing is made with an outer layer of glass and an inner layer or layers of plastic film type glazing. The recommended inner glazing is Teflon FEP film, made by Du Pont. Refer back to figure 3–15, which indicates the recommended layers of glazing for various climates. If you live in the South, your batch collector will need at least two layers of glazing. A batch collector in northern states and Canada requires three layers. You can refer to table 4–1 to review your options for glazing materials.

The first step after you've selected your glazing materials is to build the glazing frames. These frames are separate from the collector box so they can be removed if the need ever arises.

To build the two glazing frames, use #2 pine to make six pieces that measure ¾ x 1⅝ x 86⅛ inches and six pieces that measure ¾ x 1⅝ x 53⅜ inches. One of these frames will carry one or two layers of plastic glazing, and the other will serve as both a spacer and a seal for the exterior layer of glass. Referring to figure 4–13, cut the end-lap, cross-lap and T-lap joints on these boards, and arrange the pieces as shown in the figure. Lay them on a sheet of plywood, using the plywood's outer edges as a guide to square the frame. Assemble the frames with wood glue and two #6 x ¾ flathead wood screws per joint. When the

(*continued on page 98*)

TABLE 4–1
Solar Glazings

Material Type and Form	Trade Name (manu-facturer*)	Comments	Installation Suggestions	Suitable Applications† outer glazing	inner glazing	
GLASS rigid, flat sheet tempered and annealed (window glass) tempered and annealed low iron (high transmittance)	many Clearlite (AFG) Sunadex (AFG) Heliolite (Hordis Brothers) SolaKleer (General Glass) plus many others	(Some brands of low-iron glass have textured, light-diffusing transmission.) Excellent transparency and appearance Excellent resistance to UV, weather and high heat Low thermal expansion/contraction Readily available Noncombustible, chemically inert Low impact resistance; potential for breakage a safety hazard in some areas Heavy, requires strong supports Hard to handle on site	Bottom edges must be supported at quarter-points on neoprene "setting blocks." Butyl tape ("glazing tape") is a common gasket and sealant for glass.	yes	yes	
ACRYLIC rigid, flat sheet double-wall extrusion	Plexiglas G (Rohm & Haas) Lucite L, Lucite SAR (Du Pont Co.) ACRYLITE GP (CYRO) Acry-Pane (Sheffield) EXOLITE Acrylic (CYRO)	(Double-wall extrusion is semitransparent.) Good transparency and appearance Good UV and weather resistance Lightweight Readily available in many sizes and thicknesses Easy to site-fabricate High thermal expansion/contraction Susceptible to abrasion (surface scratches) Softens under moderate heat	Use mechanical compression fastening around all four edges. Mounting must be resilient to allow thermal expansion/contraction. Through-bolting or nailing not recommended. For double-wall extrusion: Special mounting hardware is available from manufacturers. Ends of open channels should be sealed.	yes	doubtful	

*See list of manufacturers and distributors in Appendix 1.
†In making these recommendations for "Suitable Applications," we have weighed the factors (namely cost, performance and long-term durability) that would be important in a typical home solar retrofitting project.
"Yes" means that the application has a very good chance of success.
"Maybe" means that the application is possible, but will require special care in design or construction to have greater than a 50/50 chance of success.
"Doubtful" means that the application might be possible, but only under very unusual combinations of cost, design or construction requirements.
"No" means that the application is almost certainly not feasible.

Table 4–1—*continued*

Typical Available Thickness and Sizes (weight, lbs./ft^2)	Typical Retail Price Range ($/ft^2)‡	Estimated Service Life (yrs.)	Solar Transmittance, % (Infrared Transmittance, %)§	Max. Continuous Service Temp. (°F)	Thermal Expansion Coefficient (10^{-5} in./ in./°F)	Impact Strength (tensile strength, psi)
⅛″ thick (1.6) ³⁄₁₆″ thick (2.5)	$0.95 to $2.75	50 + barring breakage	regular glass: 82–84 (less than 2)	400°–600°	0.47–0.51	very low (10,000– 20,000)
tempered glass commonly available in: 34″ × 76″ 34″ × 96″ 46″ × 76″ 46″ × 96″ and other sizes	up to $4.88		low-iron glass: 90–91 (less than 2)			
flat sheets: ⅛″ thick (0.78)	$1.50 to $2.00	25 +	flat sheets: 89 (less than 6)	160°–200°	3.4–4.0	medium (10,000)
¼″ thick (1.56)	$2.00 to $2.65	double-wall extrusion; 83 (less than 6)				
double-wall extrusion ⅝″ thick (1.0)	$2.50 to $3.50					

(*continued on next page*)

‡Prices shown here are for a typical 50- to 300-square-foot quantity and do not include mounting hardware, sealing gaskets, etc., which combined with shipping and handling charges can add up to 30 percent onto the glazing's stated price per square foot. Prices start to decrease drastically at quantities larger than about 500 square feet. Prices also depend greatly on the manufacturer's marketing scheme and for a given product and quantity can vary widely depending on the company from which you buy.

§"Solar Transmittance" refers to the fraction of thermal radiant energy within the solar spectrum (typically 0.4 to 2.5 microns) which is transmitted by the glazing. "IR Transmittance" pertains to the far infrared band (typically 5 to 50 microns), which represents the thermal radiation from a room-temperature (70°F) body.

TABLE 4–1—*continued*

Material Type and Form	Trade Name (manu-facturer*)	Comments	Installation Suggestions	Suitable Applications† outer glazing	inner glazing
POLY-CARBONATE rigid, flat sheet	Tuffak A (Rohm & Haas) Lexan 9030 (G.E.) Poly-Glaz (Sheffield)	(Double-wall extrusion is semi-transparent.) Good transparency and appearance High impact resistance Lightweight Questionable resistance to UV and abrasion High thermal expansion/ contraction	Use mechanical compression fastening around all four edges. Mounting must be resilient to allow thermal expansion/contraction. Through-bolting or nailing not recommended.	maybe	doubtful
double-wall extrusion (thick wall) double-wall extrusion (thin wall)	EXOLITE Poly-carbonate (CYRO) Tuffak-Twinwal (Rohm & Haas) Qualex (Structured Sheets)		For double-wall extrusion: Special mounting hardware is available from manufacturers. Ends of open channels should be sealed.		
FIBERGLASS REINFORCED POLYESTER (FRP) flexible, flat thin sheet	Sun-Lite Premium II (Solar Components Corp.) Solar G (Filon Div.) Crystalite-T (Lasco Ind.) Glasteel (Glasteel, Inc.)	Translucent, diffused light transmission Very lightweight Good impact resistance Easy installation and mounting Readily available Questionable resistance to UV, surface erosion and high heat Requires occasional surface recoating	Overlap at least ¾″ at edges of sheet onto support frame. Sheet should be pre-drilled for fasteners; holes should be at least ⅛″ oversize. Seal with silicone.	maybe	maybe
corrugated and shiplap configurations also available		High thermal expansion/ contraction Hard to eliminate wavy appearance of flat sheets Low solar transmittance at oblique incidence angles	Special mounting hardware for flat and corrugated fiberglass is available.		

TABLE 4–1—*continued*

Typical Available Thickness and Sizes (weight, lbs./ft²)	Typical Retail Price Range ($/ft²)‡	Estimated Service Life (yrs.)	Solar Transmittance, % (Infrared Transmittance, %)§	Max. Continuous Service Temp. (°F)	Thermal Expansion Coefficient (10^{-5} in./in./°F)	Impact Strength (tensile strength, psi)
flat sheets: ⅛″ thick (0.78) ¼″ thick (1.56)	$2 to $4 $4 to $6	10 to 15	flat sheets: 86 (less than 6)	200°–260°	3.3–4.0	high (9,500)
double-wall extrusion: 4, 5, 6, 7 mil. thick (0.22 to 0.31)	$1.25 to $1.50	25 +	double-wall extrusion: 74–77 (less than 6)			
0.025″ thick available widths: 24″, 36″, 48″, 49½″, 60″ available lengths: 8′, 10′, 25′, 50′ (0.25)	$0.78 to $1.03	8 to 12	0.025″ thickness: 87 (10–12; 5–50 micron band) 0.040″ thickness: 85 (5–6; 5–50 micron band)	200°	1.36	medium (10,000)
0.040″ thick other dimensions as above (0.31)	$0.96 to $1.22		0.060 thickness: 72 (less than 2)			
0.060″ thick other dimensions as above (0.50) other thicknesses: 4 oz/ft² = 0.030″ 5 oz/ft² = 0.037″	$1.29 to $1.55					

(*continued on next page*)

TABLE 4–1—continued

Material Type and Form	Trade Name (manu-facturer*)	Comments	Installation Suggestions	Suitable Applications†	
				outer glazing	inner glazing
FLUORINATED ETHYLENE PRO-PYLENE (FEP) thin film	Teflon (Du Pont Co.) Type 100A: general purpose Type C: cementable with adhesives Type L: greater flexibil-ity for environ-mental ex-tremes (to 0.090")	Totally transparent Very high solar transmittance Superior resistance to UV and high heat Low cost Very lightweight Chemically inert, noncombustible Suitable only for inner glazings High thermal expansion/ contraction (can sag at high temperatures) Not readily available in small quantities Hard to eliminate wrinkles in installation High infrared transmittance Easily torn or punctured	To prevent sag, should be stretched ap-proximately 1% during mounting, and support wires used under long spans. Use adhesives and mechan-ical clamping along edges. Nails and staples not recom-mended.	no	yes
POLYVINYL FLUORIDE (PVF) thin film	Tedlar (Du Pont Co.) Type 400SE PVF	Almost transparent Very lightweight High solar transmittance Low cost High tensile strength Questionable resistance to UV and weathering Hard to eliminate wrinkles in installation (should be shrink-mounted) High infrared transmittance Embrittlement at prolonged high temperatures Not recommended for inner glazings Unknown long-term durability Can be torn or punctured under impact	Gentle heat-shrinking prior to mounting will provide taut surface. Heat-sealing or adhesive bonding is preferred to nailing or stapling.	yes	no

TABLE 4–1—*continued*

Typical Available Thickness and Sizes (weight, lbs./ft²)	Typical Retail Price Range ($/ft²)‡	Estimated Service Life (yrs.)	Solar Transmittance, % (Infrared Transmittance, %)§	Max. Continuous Service Temp. (°F)	Thermal Expansion Coefficient (10⁻⁵ in./ in./°F)	Impact Strength (tensile strength, psi)
Type 100A (1 mil thick) 1 mil thick common. Thickness to 20 mil also available. 50″ width common. Other widths also available. 50′ length common. Up to 300′ available. (0.011)	$0.39 to $0.69	up to 20	96 (58; 3–50 micron band)	400°	9.0	very low (3,000)
4 mil thick 4″ Distributors sell Tedlar in a variety of lengths and widths (0.029)	$0.60 to $0.77	5 to 10	90 (greater than 50)	225°	2.8	low (12,000)

(*continued on next page*)

TABLE 4–1—continued

Material Type and Form	Trade Name (manu-facturer*)	Comments	Installation Suggestions	Suitable Applications† outer glazing	inner glazing
LAMINATED ACRYLIC/ POLYESTER thin film	Flexigard 7410 or 7415 (3M Co.)	Totally transparent Combines weatherability of acrylics with strength of polyester Very lightweight Relatively low infrared transmittance Good solar transmittance Does not sag at high temperatures Unknown long-term durability Hard to eliminate wrinkles in installation Nonreversible Susceptible to wind flapping Can be torn or punctured under impact	Outside of roll is side of film to be exposed.	maybe	maybe
WEATHER-ABLE POLYESTER thin film	LLumar (Martin Process-ing Inc.)	Totally transparent Very lightweight High strength for a thin film Transmits almost no UV Does not sag UV stabilizers are integral with film, not damaged by scratch or abrasion Unknown long-term durability Hard to eliminate wrinkles in installation Susceptible to wind flapping Can be torn or punctured under impact	Helpful to wrap around frame members before fastening.	maybe	maybe

TABLE 4–1—*continued*

Typical Available Thickness and Sizes (weight, lbs./ft²)	Typical Retail Price Range ($/ft²)‡	Estimated Service Life (yrs.)	Solar Transmittance, % (Infrared Transmittance, %)§	Max. Continuous Service Temp. (°F)	Thermal Expansion Coefficient (10^{-5} in./ in./°F)	Impact Strength (tensile strength, psi)
7 mil and 11 mil thick 49″ × 20′ rolls 49″ × 100′ rolls and 49″ × 200′ rolls (0.057 and 0.075)	Contact the manufacturer for current prices	7 to 10	89 (9.5; 6–50 micron band)	275°	2.7–5.0	low (14,000)
0.005″ thick available widths: from 26″ to 60″ available lengths: from 50′ to 300′ (0.03) 0.007″ thick in 48″ width and above lengths	$0.46 to $0.69 $0.64 to $0.94 both thicknesses sold by the lb. at $8.85/lb.	7 to 10	88	300°	(not available)	low (25,000)

(*continued on next page*)

TABLE 4–1—*continued*

Material Type and Form	Trade Name (manu- facturer*)	Comments	Installation Suggestions	Suitable Applications† outer glazing	inner glazing
ANTI- REFLECTIVE COATED POLYESTER thin film	SunGain (3M Co.)	Totally transparent Antistatic; will not cling or attract dust Very lightweight Unknown long-term durability Suitable only for inner glazing Relatively high cost Special surface coating easily damaged by scratch or abrasion Can be torn or punctured under impact	Surface should not be rubbed or abraded. Handle film by edges only and wear clean cloth gloves. Adhesive mounting is recommended (e.g., Scotch brand 838 weather-resistant tape).	no	maybe

SOURCE: Adapted from *Solarizing Your Present Home*, Joe Carter, ed. (Emmaus, Pa.: Rodale Press, 1981).

NOTES:

1. Prepared by David Sellers and Bob Flower of Rodale's Product Testing Department.

2. Some items in this table are based not only on data furnished by manufacturers or independent testing, but also on our subjective evaluation and judgment. This is necessary because, for characteristics such as "Service Life," manufacturers are totally unwilling to commit themselves to a guaranteed specification value.

3. CAUTION: Most plastic glazings are combustible, although in our opinion they do not pose a fire hazard any more serious than other plastic furnishings in the home (e.g., draperies, wall coverings, furniture, etc.). Some building codes may restrict or impose special conditions on the use of plastic glazings which are physically part of the building's roof or wall. Building codes generally do not apply to plastic glazings that are not part of the building (e.g., glazings on freestanding solar collectors).

TABLE 4–1—*continued*

Typical Available Thickness and Sizes (weight, lbs./ft²)	Typical Retail Price Range ($/ft²)‡	Estimated Service Life (yrs.)	Solar Transmittance, % (Infrared Transmittance, %)§	Max. Continuous Service Temp. (°F)	Thermal Expansion Coefficient (10^{-5} in./ in./°F)	Impact Strength (tensile strength, psi)
0.004″ thick 51″ × 50′ rolls and 51″ × 300′ rolls (0.029)	$0.68 to $1.00	10 +	93 (not available)	300°	3.06	low (25,000)

4. Some manufacturers produce lower quality utility-grade glazings for less demanding nonsolar applications. Examples: Plexiglas MC, Sun-Lite Regular, and Filon Types 740 and 750 (also called "Solar-Gro Home Greenhouse Panels"). Such products look (to the eye) like their higher quality counterparts, but their long-term performance will be significantly worse. Buyers should always make sure of the exact product name and type designation before placing an order to avoid being surprised by receiving the wrong material.

5. Many manufacturers don't sell small quantities of their products at retail, but they should be able to direct you to retail and wholesale sources of supply in your area.

glue is dry, round off all the sharp corners. Make sure no pointy wood screws stick out of the frames.

Lay the two frames on top of each other and drill 20 holes, $\frac{3}{16}$ inch in diameter, through both, as shown in the figure. Again sand down any rough spots made by the drill. Coat the frames with an oil-base porch paint, letting the paint dry for at least four days before giving them another coat, which should also be allowed to dry for four days.

When the paint has dried, spread some newspaper out on the floor and lay one of the frames on it. Cover half the frame with the Teflon, placing one edge on the 53⅜-inch-long middle cross-member and making sure that the other three edges overhang the frame by about 4 inches. The glazing will be stapled to the frame, and it's a good idea to reinforce all stapling points with standard glass-fiber packing tape that has been adhered to the Teflon. Starting at the cross-lap joint, staple the glazing onto the middle cross-member, using two ¼-inch staples. Then gently pull the Teflon taut and staple it at the middle cross-member. The staples should be driven 2 to 3 inches apart and parallel with the length of the cross-member. Don't staple the 86¹⁄₁₆-inch cross-member.

Turn the frame over and pull the glazing around the frame at the place where the 86¹⁄₁₆-inch cross-member meets the 53⅜-inch end-member. Gently pull the Teflon taut, and staple the glazing at the T-lap joint.

Photo 4–7: Plastic glazing film is stretched like a drumskin across the underside of the glazing frame, then stapled onto the top side. This method makes a wrinkle-free, air-tight seal.

Pull the glazing around one of the corners and staple it to the end-lap joint. Staple down the other corner the same way. Now you can complete the job by pulling the glazing edges very snugly around the frame and stapling every 2 or 3 inches. Trim off excess glazing with a utility knife or scissors.

Cover the other half of the frame the same way, but before laying the glazing down on the frame, cover the 53⅜-inch middle cross-member with a bead of clear silicone caulk. Then staple through the caulk.

One of the glazing layers is now complete. If you need just two layers of glazing, you can now set aside the glazing frame and begin installing the collector. If you're putting in a third layer (one of glass, two of plastic), turn the Teflon-covered glazing frame over so that the Teflon faces down. Then cover the other side of the frame with Teflon to create a double layer of glazing on a single frame. This side of the frame is covered in the same way you covered the first side, except that silicone caulk is applied around the entire perimeter of the frame before the Teflon is unrolled. Staple through the caulk. The silicone caulk will seal the airspace between the two layers of film.

Set the glazing frame aside. The second frame you made is a spacer that will be used to fasten the glass layer in a later step, after the collector box has been installed.

Installing the Box

What you now have is the makings of a basic box for a batch collector. The next steps involve the installation, and because every house is different there is no single set of installation steps that would work for every application. The basic box can be mounted against the wall of a house, or it can be free-standing. If one of your home's walls faces south it might be best to mount the collector against the wall, tilted at an angle equal to your latitude. That way little or no connecting plumbing will be exposed to freezing temperatures. If your home's walls don't face south, or if it's not convenient to install the collector against a wall, a freestanding mount might be best. A freestanding batch collector should still be installed as close to the house as possible to minimize the lengths of the two pipe runs.

Support for the box can be gotten in a number of ways. Photos 4–8 through 4–11 show how one homeowner made a three-sided concrete foundation to support a tilted batch collector that is built right up to the house. The best way to get a good visual integration with a wall-mounted system is to match siding and colors as best you can. A concrete slab would work in an installation like this. In the cold climate where this installation was done, the foundation extends below the frost line. The collector was mounted on the foundation with pressure-treated 2 x 4's that are fastened with standard anchor bolts. The finished collector is actually not nailed to the side of the house, but rests neatly alongside the wall. This is to prevent damage to the house in winter because when the ground freezes, the foundation may be lifted by freezing water in the soil. The collector could rip loose if it were nailed to the house. This is an important rule: In very cold climates don't attach a ground-supported collector directly to the side of the house.

Along with being fastened to the foundation, the collector must also be stabilized against wind and snow loads. The freestanding installation in photos 4–12 through 4–14 shows a concrete footing and a simple arrangement of cross-braced 2 x 4's.

Whichever way you create a mount for the collector, all exposed wood should be chosen to last a long time. Redwood and ce-

Photos 4–8, 4–9, 4–10,4–11: A three-sided concrete footing can carry the entire weight of the batch. First form boards are set in place, above left; then the concrete is shoveled in. Anchor bolts, spaced evenly around the footings, are set in the concrete while it is still wet. After the concrete has hardened, 2 x 4's are bolted around the perimeter of the foundation, above right. A penetration is made through the wall of the house, and then the collector is set in place. After it has been tilted to the proper angle, it is braced in the back with 2 x 4's, right, which are nailed with truss plates to the 2 x 4 sills and the collector's long side frame studs. The collector is never nailed to the wall of the house, as the concrete footing can shift when the ground freezes and thaws. The bottom of the collector is braced with 2 x 4's and truss plates. The hot and cold water pipes are wrapped with heat tape, insulation, and PVC pipe sleeves, then passed through the wall penetration. After several more 2 x 4 side and bottom braces have been added, opposite, the collector is ready for the finish siding.

dar, though relatively expensive, are much more rot resistant than other commonly available wood. Pressure-treated wood is also a good bet for durability, and plain old paint (or Cuprinol or Penta-Seal) can't be denied for its preservative ability.

We'll continue describing the ground and ground/wall mount in more detail, since it's a popular choice for batch systems. After you've decided where you'll install the batch collector, pour the foundation (in cold climates) and wait for the concrete to dry. Don't forget to insert about a dozen 8- or 10-inch anchor bolts around the perimeter of the foundation while the concrete is still wet. An installation in milder climates can use concrete piers instead of the continuous foundation, or some other method of keeping the box off the ground.

To get the two pipe runs from the collector into the house you must cut an access hole through the wall. The hole should be as close to the collector as possible to minimize the distance of exterior pipe runs. Remember, you must connect the collector to your conventional water heater, so the hole you cut should also be as close as possible to the water heater. (It's also very important to have all outside plumbing lines below the average frost line depth in your area.) This often means cutting a hole through the foundation to get access to the basement. Before you cut, make sure the hole won't interfere with electrical wiring, existing pipes or floor joists.

A hole through concrete or concrete block can be chiseled by hand or cut with an electric or air-powered chisel hammer. Once the hole has been made, insert a 4-inch-diameter piece of PVC pipe to serve as a sleeve for the water pipes. The pipe should be long enough to protrude a few inches from both sides of the wall. Angle the pipe or locate the hole so the outside end is 3 to 4 inches above ground level, and then support it so that it stays in this position. Then use mortar cement to seal the remaining gap around the sleeve.

Returning to the collector box, attach the four tank support brackets inside the reflector cavity using ⅜-inch lag bolts and washers. Then install the tank and its supports using ⅜–16 x ¾-inch hex head bolts, nuts and washers.

Stand the collector upright. Assemble the pressure-and-temperature relief valve and drainpipe according to the directions in figure 4–12. Attach the valve to the tank and find the proper place to drill a hole through the back of the collector to accommodate the drain tube (figure 4–11 for the suggested place to drill this hole). Drill a ⅝-inch hole through the Mylar-covered reflector cavity and the plywood back of the collector. Connect one

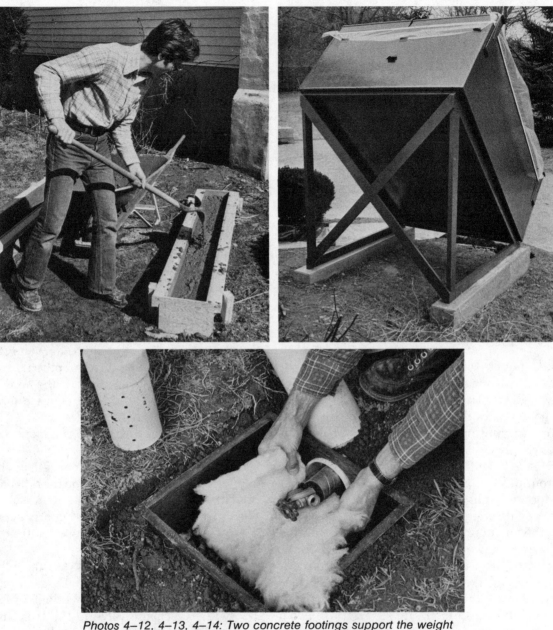

Photos 4–12, 4–13, 4–14: Two concrete footings support the weight of a freestanding batch collector. Form boards are made; the concrete is poured, and anchor bolts are set into the wet concrete. When the concrete has dried, a 2 x 4 is bolted to the top of each of the footings. The collector is set in place, tilted to the proper angle, then securely cross-braced with 2 x 4's in the back. The front can be nailed in place with truss plates. In this installation the tank drain valve was hidden underground. A hole was dug, a wooden box was made, and a valve was wrapped in insulation. Later a cap was made from a 4-inch-thick PVC drainpipe and slipped around the insulation, and the entire drain cavity was covered with a removable concrete paver.

end of the drainage tube to the pressure-and-temperature relief valve and pass the other end through this hole.

With the foundation and the mounting assembly built, you can at this stage put the nearly finished collector into its place. Nail a 1½ by 3½ by 96-inch board to the collector's back with 12d common nails. This is a temporary lifting board. You'll need help from three or four people to lift the collector onto the mounting assembly. Remove the lifting board when the collector has been positioned properly, and fasten the box to the mounting assembly with 1½ by 6-inch truss plates and 1-inch underlayment nails.

Study the schematic in figure 4–14 to see how the collector should be hooked up to your home's plumbing. Seven valves attached near the conventional water heater will allow for three different modes. Most of the year the batch system will work as a preheater. When hot water is used, the cold replacement water will be diverted by valves #2 (closed) and #3 (open) to the collector. Heated water then flows from the batch collector to the conventional water heater when there is a hot water draw. This replacement water is called solar "preheated" water because it may not have reached the full hot water temperature. However, the water heater will use much less energy to boost the temperature of the preheated water (which can get as hot as 100°F) to the required 110° to 115°F than it would heating water up from 50°F. In the late spring, summer and early fall valve #5 can be opened and valve #4 closed, and the conventional water heater can be shut down completely so that your hot water will be solar-heated only. This is a mode you'll have to experiment with to make sure that your load doesn't exceed the amount of hot water the batch system can supply. Also, the "when" of shutting down the conventional water

heater is another variable. If you do it too early in the spring or leave it off too far into the fall, there may not be enough solar energy at those times to supply 100 percent of your hot water needs. If you decide you must drain the collector in the wintertime instead of risking a freeze problem, open valves #2, #6 and #7 and close valves #3, #4 and #5. This decision is a little trickier because you certainly don't want to wait to see if the collector will, in fact, freeze. A little later in this chapter there will be some explanation of how to minimize the freeze vulnerability of your batch system, enough so that in most cold climates you'll be able to operate it through the winter.

It was explained earlier that in freezing climates ground-mounted collectors should not be attached to the side of the house. Freezing ground could make the foundation heave, raising the collector and ripping it loose. For the same reason, the collector should be connected to the home's plumbing with ½-inch-i.d. soft copper tubing, which will bend and give if the ground heaves a little. You can also use CPVC or polybutylene tubing for this connection.

You should preplan the route by which you'll run the tubing to connect the collector to the water heater. If the collector is against the wall, you can run the tubing through the hole in the wall straight to the collector. If the collector stands more than several feet away from the house, you'll have to dig a shallow trench to contain the tubing. Underground tubing runs should be well insulated and covered by PVC pipe. Some brands of pipe insulation are made with foam-filled PVC pipe sections that are ready for tubing insertion (see Appendix 1).

Attach the cold water inlet and hot water outlet tubing to the collector's water tank, as shown in figure 4–12. Use a tubing bender to bend the tube to fit under the collector.

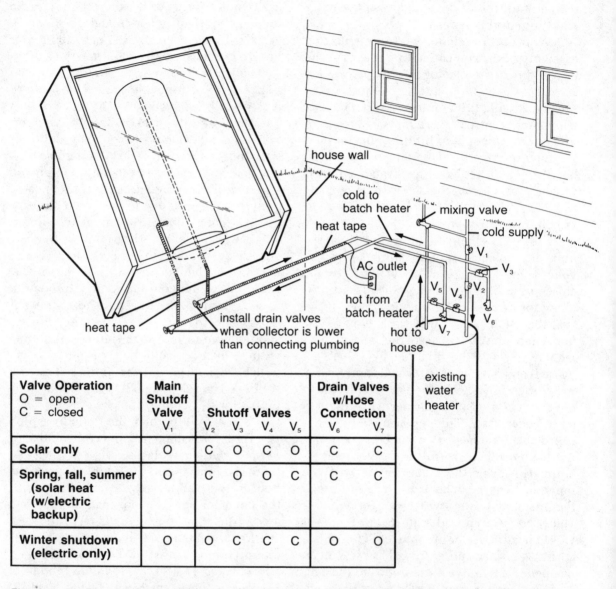

Valve Operation O = open C = closed	Main Shutoff Valve V_1	Shutoff Valves				Drain Valves w/Hose Connection	
		V_2	V_3	V_4	V_5	V_6	V_7
Solar only	O	C	O	C	O	C	C
Spring, fall, summer (solar heat (w/electric backup)	O	C	O	O	C	C	C
Winter shutdown (electric only)	O	O	C	C	C	O	O

Figure 4–14: By adding valves to the plumbing near your water heater you'll be able to quickly bring the batch collector in and out of service. You can make a copy of the valve chart and place it in a conspicuous spot near the water heater.

The tubing bender can also be used to shape the tubing to fit close to the water tank. Make sure to use all the unions and other fittings shown in figure 4–12. Don't forget to install the vacuum breaker at the top of the tank. This will let air into the tank when it's drained and make the draining go much faster.

Run the inlet and outlet tubing through the hole in the wall and connect them to the water heater as shown in figure 4–14. Use plumber's strap every few feet to secure the pipe runs to the ceiling or walls. The inside joints are sweat soldered (see figure 1–3 for sweat soldering instructions). Don't forget to add the mixing valve somewhere along the outlet line. This prevents scalds and conserves solar-heated water by automatically mixing in cold water when the collector outlet water is too hot (above 125°F). Valves #6 and #7 are drain spigots (hose bibs) that should be connected to hoses. The hoses should lead to a drain. When a basement water heater is connected to a collector at a higher elevation (such as out on the lawn), drain valves #6 and #7 should be installed at the lowest point in the connecting plumbing to allow total draining of all exposed plumbing in the system.

Be sure to shut off the water heater's electricity or gas supply before connecting the batch heater. And before cutting into your home's existing plumbing, you'll have to close the main water valve.

Once all the plumbing has been connected, open the appropriate valves and fill the batch heater. You should pressure-test all the new plumbing for 24 hours by allowing cold water to pass through the collector before glazing the collector and insulating the water pipes. If you want, you can use a pressure gauge to check for leaks. When installed in the pressure test "loop," it will tell you if there are even slight leaks by indicating a drop from the original pressure reading when the system was filled. When you're certain there are no leaks, wrap the exterior pipe runs with protective heat tape. The recommended type is continuously self-regulating heat tape (Frostex II, FreezGard, both made by the Raychem Corporation; see Appendix 1) which uses no switching-type thermostat. Instead the heat tape itself responds to temperature and literally sends heat (electric current) only to points along a pipe run that are excessively cold. This tape is simply plugged into an outlet (120 VAC) on one end. All piping that is outdoors and above ground should be wrapped with the tape, even the pipe inside the collector box. Then all the "taped" pipe should, of course, be insulated.

After the pipes have been protected, you can finish insulating the collector. Cover the bottom of the collector with 3½-inch fiberglass insulation. Drive a few nails into the studs surrounding the insulation and run wire between the nails to hold the insulation in place.

Now install the glazing frame. Position the frame on the collector box and caulk around the top side. Lay the spacer frame on top of the caulking and the glazing frame, and fasten them down with about twenty #8 x 2½-inch flathead wood screws.

Cover the top side of the spacer frame with glazing tape. This tape is made of soft butyl rubber that has a very strong adhesive ability. Then install the glass. The box dimensions are too big to be enclosed with a single pane of glass, so you'll need to buy two panes that have each been cut to 51½ by 42 inches. Install the lower pane of glass first by placing it on the glazing tape. Put the second pane on, leaving about a ⅛-inch gap between the two edges of the middle, then use

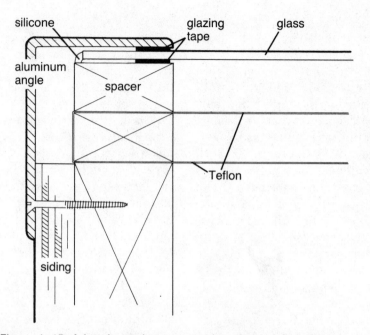

Figure 4–15: A batch can have two or three layers of glazing, depending on the climate (see figure 3–15). The bottom layer or layers can be a plastic film, such as Teflon. To protect the plastic, the top layer should be glass. The film is attached to the glazing frame with tape and staples. The spacer frame is placed atop the glazing frame and then the glass is installed with glazing tape and silicone caulk. Everything is held down with screws and aluminum angle. This illustration depicts triple glazing: two layers of film topped by a layer of glass. If your region requires double glazing, eliminate one film layer.

silicone caulk between the edges of the glass and the wood and fasten it all down with 2½ by 1-inch aluminum angle and 1-inch wood screws, as shown in figure 4–15. The face of the angle that touches the glass is lined with more glazing tape. The wide face of the angle should be predrilled, and when screwing it to the wood, press down on the angle to get a good compression.

Cover the exterior sides of the collector with AC-grade exterior plywood or paneling. Some people spend almost as much time building a custom enclosure as building the collector itself. Only you can decide how to make the collector look attractive in your yard.

Many homeowners prefer to cover the collector with the same siding and trim found on the house. They want to make the collector blend in with its environment. Suppose, for example, that your house has white clapboard or aluminum siding and black window shutters. The sides of the collector can be covered with the same siding, and the insulating doors painted black to resemble the shutters. Even more exotic siding materials, like cedar shingles, can be used to cover the collector enclosure. Freestanding collectors are sometimes trickier to finish because they have such an unusual shape. You might want to disguise a freestanding unit as a tool shed

Photo 4–15: The completed glazing frame is slid into place to rest on the framing of the box. The frame is then secured with aluminum angle or prebent flashing.

by building storage space onto the back as suggested in figure 4–16. However it's done, the right materials and colors will help to integrate this big collector with its surroundings, and, with time, it won't look so big anymore.

Freeze Protection

If you want to further decrease the freeze vulnerability of your collector, you still have the option of building insulating doors. They will also improve the performance of the batch system, no matter what the weather, because the doors will help to minimize heat loss from the batch tank. Climate conditions are part of the decision here, but so is your commitment to the daily tasks of opening and closing the doors.

The doors themselves are fairly simple. Start by cutting four 86-inch pieces of 1 x 4 pressure-treated lumber and four 28¼-inch pieces. Rip the pieces down to a width of 3 inches. Nail these pieces together as shown in figure 4–17. Then cut four pieces of ¼-inch AC-grade exterior plywood to 30¼ by 86 inches. Lay one of the door frames down on a flat surface, place one of the plywood pieces on the door frame and nail it down using 1-inch underlayment nails. Turn the frame over and fill the 3-inch-deep cavity with three layers of 1-inch extruded polystyrene

Photo 4–16: This collector's paneling and flashing neatly match the house. Notice the shingles at the top of the collector.

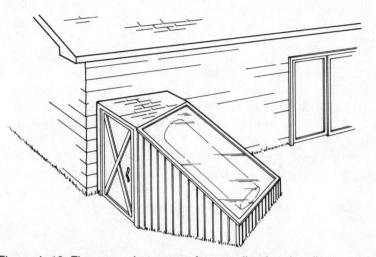

Figure 4–16: The space between a freestanding batch collector and the wall of a house can be used for storage. Close in the space and you've got a storage shed.

insulation (e.g., Styrofoam). Enclose the insulation by nailing the second plywood section to the door frame. Make the second door the same way and then paint them. Add handles and attach the doors to the collector with large hinges. As shown in figure 4–17 the hinges were attached to 16-inch-long boards that were in turn nailed to the side of the collector. Once the doors have been securely installed, add weather stripping to the inside edges of the doors. Now all you need is solar energy. A final enhancement of these doors would be to make them double as reflectors when they are in the open position. The inside face can be covered with aluminized Mylar or other reflective surface, and the open position of the doors should be such that reflected sunlight bounces into the collector box.

After the system is working, it's a good idea to copy the valving chart in figure 4–14 and tape it near the water heater. The diagram will serve as a guide to the proper seasonal valving positions, and if you ever move

out of the house, the next occupant will know what the system is all about.

The batch heater is simple to operate and maintain. Every morning you must open the insulated doors to expose the water tank to the sun. In the evening, when you get home, close the doors to keep the heat through the night. The schedule of seasonal valving changes for full or partial solar heating is something you'll have to work out in the first year of operation.

An important maintenance task involves testing the heat tape by dipping the thermostat into ice water. If the system is not drained during the freezing months, you should try to check the heat tape at least every other month. You can also check the plumbing for tiny leaks and the collector in general for the condition of paint, seals, glazing and so forth. The glazing will probably need to be cleaned periodically. If you do decide to drain the collector over the winter, you simply have to close the valve on the collector inlet, open the drain valves and unplug the heat tape.

Variations

We've said it before, and we'll say it one last time: The ultimate location and design of the installation of a batch system is highly site-specific. The box-on-the-ground and box-on-the-wall approaches aren't the only ones, so if you don't have those options, you're not necessarily restricted from using a batch collector.

Another possible location is actually higher up on a wall, whereby the collector also becomes a sort of shading overhang, as shown in photo 4–17. The triangular collector box is mostly constructed on the ground and then fastened to the studs or masonry of the existing house wall. The inside of the collector is lined with aluminized Mylar, although it's not a formal cusp reflector. In this collector you could build a larger box to enclose two tanks plumbed in series (figure 4–18). The increased glazing area and doubled tank volume would make the system somewhat more powerful. When the collector is con-

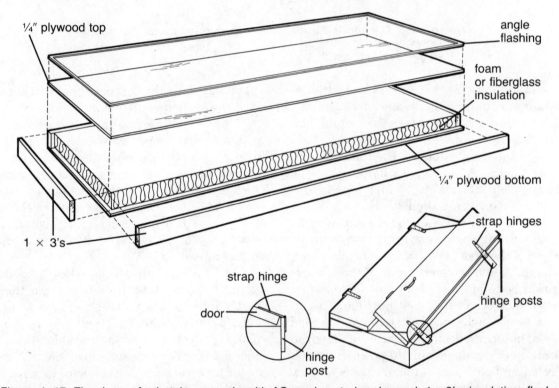

Figure 4–17: The doors of a batch are made with AC-grade exterior plywood, 1 x 3's, insulation, flashing and weather stripping. Make the 1 x 3 frame first, nail one sheet of plywood in place, fill the cavity with fiberglass insulation, and nail on the back sheet of plywood. Make the other door the same way. Finish the outside edges with flashing. The inside edges that come in contact with the glazing should be weather-stripped. Two hinge posts are nailed to each side of the collector, and the doors are fastened to these with door hinges. Paint the doors with a paint or wood preservative and add handles to make opening and closing easy.

Photo 4–17: This overhang batch collector was part of the house's original design. Recently, more and more new homes are being built with integral solar water heaters.

nected directly to the house, freeze protection is simplified. You can make a small vent (1 square foot) with an insulated door to connect the heated living space with the collector enclosure. On winter nights, open the vent to allow room air to maintain above-freezing temperatures inside the collector.

You can also consider installing the collector into your roof, as if it were a skylight. Figure 4–19 gives the details of that job. After cutting away shingles and sheathing you'll probably have to cut through one or two roof rafters and use headers to tie the cut ends

into the uncut rafters on either side of the opening. One challenge is getting the collector box, tank and assorted plumbing up on the roof. If you don't happen to have a crane, it would be best to install the box first and then go up with the tank and other hardware. Since the collector is directly connected to the house, you can make a freeze protection vent like the one suggested for the overhang collector.

Caution: If you're going on vacation for several days in warm weather, it is best to shade the collector to prevent overheating and

Photo 4–18: A batch collector installed in a greenhouse doesn't need much insulation, just direct sunlight. This collector is little more than a black tank encased in a single layer of plastic glazing. Wood and angle iron brace it securely against the wall.

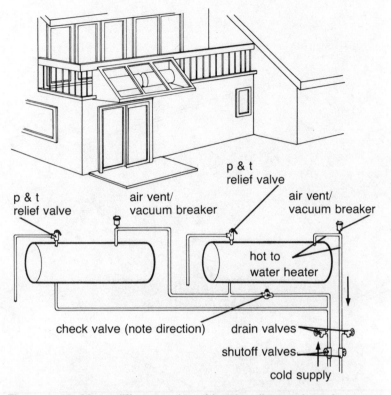

Figure 4–18: Many different styles of batch collectors have been successfully built. Here we see an overhang batch collector with two water tanks plumbed in series. If you decide to build an overhang collector like this, make sure it's supported adequately. A single 40-gallon water tank, when filled, weighs 400 pounds.

possible boiling of the water. You should either close the insulated doors (if you have them) or cover the glazing with a tarpaulin or shade cloth. Even partial shading, such as from bamboo-slat shades, is effective.

How Well Will It Work?

A batch collector like the one described here was constructed and monitored for *New Shelter* magazine. At the same time, four flat plate collector systems—thermosiphon, draindown, drainback and phase change—were also monitored. The cusp reflector batch collector was found to cost, on the average, about half as much to build and install as the flat plate system, yet delivered about 80 percent as much energy.

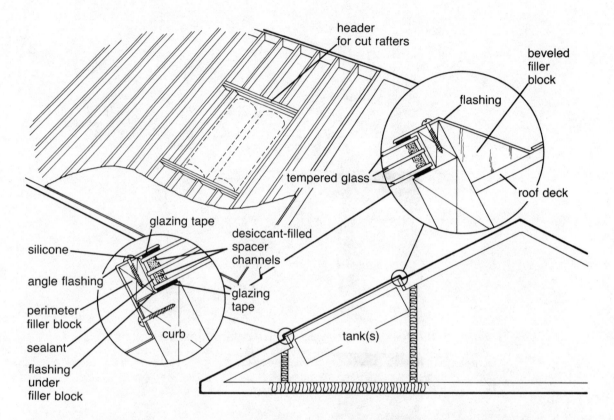

header
for cut rafters

beveled
filler
block

flashing

tempered glass

roof deck

glazing tape

silicone

desiccant-filled
spacer
channels

angle flashing

glazing
tape

perimeter
filler block

curb

sealant

tank(s)

flashing
under
filler block

Figure 4–19: The main concerns with a roof-integrated batch system are with the roof's strength and with properly sealing and flashing the collector to the roof. In a retrofit it will probably be necessary to cut some rafters and possibly even double-up some to carry the weight of one or two water tanks. Flashing and sealing is done in standard fashion, much like it is with a site-built skylight.

Photo 4–19: If you find that a roof is the best place to put your batch, you build it into the pitch to make it less of a lump. This installation is being done on a new house, but retrofitting is also possible. See the construction details in figure 4–19.

5

BUILDING AND INSTALLING FLAT PLATE COLLECTORS

The first flat plate collector was patented in 1909 by William J. Bailey, a California engineer. It was part of a thermosiphon system that included an insulated water tank inside the house. The inventor called his passive system the "Day and Night Solar Heater." Before Bailey's system came on the market, several manufacturers had been selling batch collectors. The advantage of the Day and Night system was that the insulated water tank inside the house held heat much better than a batch collector's exterior water tank. Bailey's salesman pointed out that the owners of batch heaters often had to wait until midafternoon for their water to heat up, while owners of Day and Night systems could draw hot water first thing in the morning and on into the night—"day and night."

The Day and Night Solar Heater was not an immediate success because at $180 it was considerably more expensive than batch heaters then being sold. But the luxury of 'round-the-clock hot water soon won out, and the Day and Night Company became the dominant solar water heater manufacturer in southern California.

Unfortunately, the company was almost bankrupted by a record-breaking cold snap in 1913. Collectors froze and tubing burst throughout the region. Bailey went back to the drawing board and designed the world's first freeze-protected flat plate system. The new flat plates were filled not just with water, but with a mixture of alcohol and water. The solar-heated antifreeze solution was passed through a heat exchanger in the storage tank, which heated the domestic water. The system worked and was the precursor of today's antifreeze fluid systems.

The Day and Night Solar Heater Company flourished for many years. Its product was so simple that a trade magazine observed in 1914 that "the man who was handy with tools and pipe wrenches could build his own." The demand for the system didn't ebb until the 1920s, when large amounts of natural gas were found in California, making it cheaper to heat water with gas than with the sun.

Though the Day and Night Solar Heater Company has long since gone, flat plate collectors have retained many of the advantages and disadvantages that they had during its time. Batch systems still are less expensive, but more people nevertheless own flat plate systems because of their convenience. As it was in Bailey's day, freeze protection remains extremely important for flat plates, even those installed in sunny southern California. A number of different approaches to freeze protection have evolved since Bailey mixed alcohol and water, and today's flat plate systems enjoy high reliability through the coldest weather. Perhaps the nicest thing about flat plate systems is that they really can be built by anyone who is handy with tools and pipe wrenches. It's not too hard to build a flat plate system that performs as well as much more expensive store-bought systems. If your skills include some basic woodworking, plumbing and wiring, you're certainly a candidate for the rank of solar do-it-yourselfer.

A Home-Built Collector

This chapter explains how you can build and install a simple flat plate collector, or two, three or more. The collector described here is a direct descendant of the one pat-ented by Bailey in 1909, and it is meant to be used with one of the systems discussed in the next chapter.

Depending on your location (figure 3–15) this collector can be built with one or two layers of glazing. To review, collectors in southern climates need only one glazing layer, while in the north two layers are needed.

The collector described here can use any of the commonly available absorber plates, which come in a variety of sizes. This design is flexible because it's made of wood and plywood and can be custom sized. For convenience in handling, though, it's recom-

Photo 5–1: This is a photograph of a thermosiphon flat plate system, circa late 1930s. The storage tank was hidden from view in the fake chimney.

mended that the absorber plate you buy be no wider than 3 feet and no longer than 10 feet. Big collectors are heavy and hard to install.

That brings us to the absorber. Everything in this collector but the absorber is homemade for the simple reason that it's likely to be worth your time and effort to fabricate something that is better made in the factory. You won't save much money, and it's doubtful that you can compete with today's advanced manufacturing techniques.

What you must do, is find yourself a ready-made, which may or may not be easy, depending on where you live. (Some sources are listed in Appendix 1). You're looking for quality in an absorber plate: The plate itself should be copper, and so should the water tubes, which should be soldered to the plate, not just fastened or bonded. Minimum tube size for the risers (figure 5–1) is ⅜ inch i.d. for pumped systems and ½ inch i.d. for a thermosiphon system. The header should be at least ½ inch i.d. (¾ inch is preferable) for

pumped systems and ¾ to 1 inch for thermosiphon flow. Thermosiphon systems need bigger pipes in order to minimize the effect of friction between the water and the wall of the pipe. Friction reduces flow, and reduced flow means lower efficiency. The forced flow of pumped systems is much less sensitive to friction effects.

The absorber you end up buying provides the key measurements for building the rest of this collector. Your first step toward determining the lengths and widths of the other pieces is to measure the length and width of the absorber as shown in figure 5–1. Add at least ½ inch, or more, to each measurement to "round off" to a whole number of inches. For example, if your absorber measures 21⅜ by 93½ inches, add ⅝ inch to the width and ½ inch to the length to get a *working size* of 22 by 94 inches. From this working size you can figure out the dimensions of the back, the frame and so forth. We can't give you all possible dimensions for all possible absorber sizes, so for this project we're using an ab-

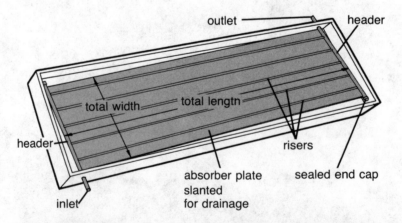

Figure 5–1: A copper absorber plate with headers and risers, a well-insulated snug box with glazing, that's what flat plate collectors are made of. The headers bring fluid into and out of the collector box, and the risers carry the fluid across the solar-heated absorber plate.

sorber plate with, you guessed it, a working size of 22 by 94 inches. (This will make a collector with about 14.3 square feet of surface area.) To repeat, all further measurements will come from these dimensions (see the materials list for materials required).

Collector Construction Steps

Make the back of the collector first. Cut a ⅜-inch piece of exterior-grade AC plywood to 24 by 96 inches. This is a half sheet, which means you can build two collectors from a full sheet. If your absorber is longer than 8 feet, you can buy a 10-foot-long sheet or join a short piece of plywood or solid wood with the longer piece (figure 5–2). This approach might save the somewhat higher cost of extra-long plywood.

The collector frame is made from good-quality 1 x 4's that have been kiln dried. The wood should be as knot free as possible. It's best to use durable woods like cedar or redwood, but you can also use standard pine (#1 or #2 grade) if it has been painted or coated with a preservative. The two sides of the frame should be cut to a length of 96 inches. Cut the ends to 24¾ inches. Check the edges of these frame pieces for bowing. Ideally, the pieces should be straight, but if an edge is slightly bowed up, use that edge (the *crown*) for the top side of the collector frame.

The next step is to fit everything together. The bottom edges of the four frame pieces must be rabbeted with ⅜ by ⅜-inch grooves to receive the plywood back. Then cut ¾ by ⅜-inch rabbets at the ends of the two end pieces (figure 5–2). Now you can join the frame pieces together with #8 x 1½-inch stainless steel wood screws. (The variable in this operation is in the type of absorber plate you're using. Some absorbers have inlet and outlet pipes on the sides, while others are

plumbed at the ends. Depending on the type of absorber, you'll have to leave either an end or a side piece unattached until the plate is installed in the box.) Countersink the screw holes and use two screws per joint. After you've driven the screws, seal the corner joints and exposed end-grain with silicone caulk, and caulk the countersunk screw holes, too.

Attach the plywood back to the frame with ring-shank nails. Use silicone along the joint where the plywood meets the frame,

---

Materials Checklist

This is a materials checklist for single- and double-glazed flat plate collectors. Dimensions, of course, will vary, depending on the size of the absorber plate.

All-copper absorber plate
1" x 4" clear, kiln-dried redwood, cedar or pressure-treated pine
#8 x 1½" brass or stainless steel flathead wood screws
1 sheet ⅜" CDX plywood
1¼" ring-shank nails
1" polyisocyanurate rigid foam insulation
Glass or plastic glazing
Metal (such as aluminum angle) or wood corner trim
Metal or wood flat trim (for glazing center break)
#8 x ¾" panhead aluminum screws (for aluminum corner trim)
Flat black paint
Wood sealer/preservative or paint
Silicone caulk
Butyl glazing tape

Additional materials for double-glazed collectors:

¾ x 1" wood sticks for glazing frame
Glazing film
Packing or duct tape
#6 x 1½" flathead wood screws
Closed-cell foam weather stripping
1¼" staples

and smooth the caulk into the joint. Then turn the box over.

Cut a 1-inch-thick piece of foil-faced polyisocyanurate insulation (brand names: Thermax, R-Max) to approximately 23¼ by 95¼ inches, so that it fits snugly inside the box. It is recommended that you cover the cut edges of the foam with aluminum tape to prevent outgassing that could deposit a film on the inside of the glazing, thereby reducing transmittance. It's also a good idea to cover the remaining exposed wood surfaces inside the box with tape.

Next cut four absorber support sticks from ¾ by ½-inch strips of pine. The support sticks should be as long as the absorber is wide (about 22 inches; see figure 5–3.) Space them at equal intervals inside the box and fasten them with #6 x 1¾-inch wood screws. Drive the screws through the back of the box, through the insulation and into the support sticks.

Once you have fastened the support sticks, you can paint the inside perimeter of the box with flat black paint. This is done for good looks, to cover the foil-faced insulation and make the inside of the collector uniformly black.

When the paint has dried, lay the absorber into the box, slanting the plate slightly so that when the collector is installed square

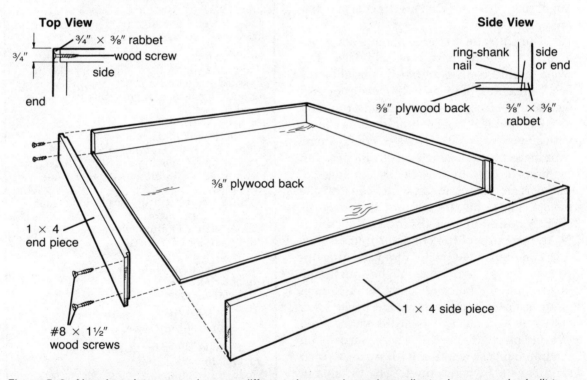

Figure 5–2: Absorber plates come in many different sizes, and wooden collector boxes can be built to fit them all. The box should be long and wide enough to hold the absorber after it has been tilted slightly for drainage. Rabbets are cut into the bottom inside edges of the 1 × 4's to receive the plywood back. The ends of the end pieces are also rabbeted to form a stronger joint with the side pieces.

Side View

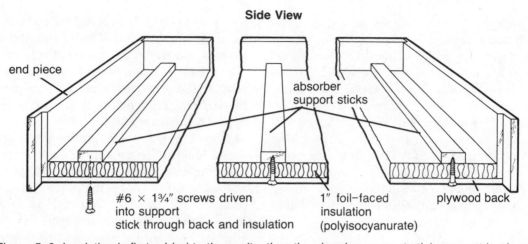

end piece

absorber
support sticks

#6 × 1¾" screws driven
into support
stick through back and insulation

1" foil-faced
insulation
(polyisocyanurate)

plywood back

Figure 5–3: Insulation is first added to the cavity, then the absorber support sticks are set in place. Three or four support sticks are usually adequate.

Photo 5–2: The glazing is carefully lowered into place.

to the roofline, it can drain more easily because the absorber is slanted (figure 5–1). Mark the points where the inlet and outlet pipes will pass through the frame. Then drill the holes to the proper diameter.

After the holes have been drilled and the absorber laid in place, attach the fourth collector frame piece to the frame and the plywood back. Caulk the joints and screw holes as before.

Then wrap the box in a polyethylene plastic sheet so that the collector is completely covered and set it outside in the sun for a couple of days. This exposure will bake out any residual wood or paint vapors that could deposit a film on the inside of the glazing.

Now it's time to add the glazing. Single-layer glazing will be discussed first. Collectors up to 9 feet long can be covered with one piece of ³⁄₁₆-inch glass. If your collector is longer than 9 feet you'll have to use two panes. With two-piece glazing, the length of each pane should be ¼ inch less than half the length of the collector. This will allow a

¼-inch gap between the two panes, and a ⅛-inch inset along the edge of the collector. Similarly, the width of the glass should be ¼ inch less than the width of the collector, to provide a ⅛-inch inset around the entire perimeter.

Collectors with two panes of glass will need a glazing center support. The glazing center support should be 1 inch wide and 24¾ inches long. Its thickness should equal the distance between the top of the absorber and the top edge of the collector frame (figure 5–4), so that it will not only serve as a glazing support, but will also help to hold down the absorber. The glazing center support should be notched as shown in figure 5–4. Cut corresponding 1 by ½-inch notches in the exact middle of the length of the collector. The center support should fit snugly in the notches and it should rest on the absorber. Screw the glazing center support in place.

Thoroughly clean the glass and vacuum the collector. Lay down strips of butyl glazing tape on the edge of the collector frame, with two strips along the glazing center support, if there is one. Then lay the glass in place. Using silicone, caulk around the seam between the glass and the wood. Cut lengths of 1 by 1½-inch aluminum angle to appropriate lengths. These will be used as glazing caps. Add strips of glazing tape to the face of the aluminum angle that touches the glass (figure 5–4). To get a good seal get a helper to press the glazing caps onto the glass while you drill the guide holes and drive the pan-head screws (#6 x ¾-inch) into the sides of the collector frame.

The final step, if it's needed, is to make the center strip. It should be 1 to 1½ inches wide and as long as the distance between the glazing caps. Apply two strips of glazing tape along the bottom side of the center strip and

Collector Cross Section

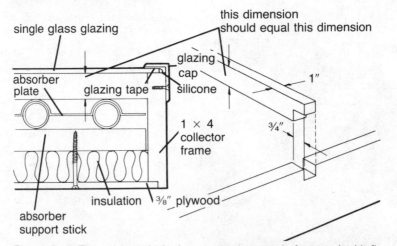

Figure 5–4: The rabbeted glazing center support is fastened with flat-head screws into the notches cut in the center of the collector side pieces. The top of the center support should be flush with the top of the side piece, and the center support should be deep enough to hold down the absorber plate. A single layer of glazing is fastened down with glazing tape, silicone caulk and aluminum angle (glazing cap).

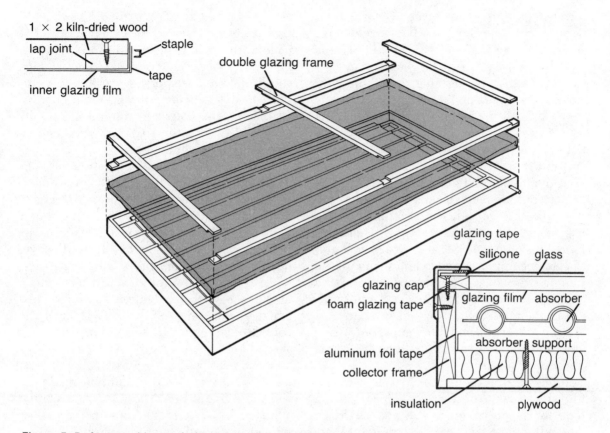

Figure 5–5: A second layer of glazing, in this case plastic film, is held down with a double-glazing frame. The glazing film is fastened to the frame with tape and staples. The frame is held down with screws, driven from above through the plastic. The outer layer of glazing—always glass—is held down with silicone caulk and aluminum angle. Glass can also be used for the inner layer, though you won't be able to drive screws through the glazing frame.

center it over the gap between the two panes, above the glazing center support. Attach the predrilled center strip to the glazing center support with screws, driving the screws through the ¼-inch gap between the panes. Be very careful not to touch the glass with the screws, as the slightest pressure on the edge can crack the pane.

Paint all the wooden surfaces with oil-base paint or a high-quality wood sealer/preservative. Then apply silicone around the places where the inlet and outlet pipes pro-

trude through the collector box, and you're done.

Double Glazing

Double glazing requires a little more work. A glazing frame assembly must be made from ¾ by 1-inch sticks to the exact outside dimensions of the collector frame, as shown in figure 5–5. The assembly includes a center divider. Lap joints are cut as shown in the illustration, and the assembly is held together with countersunk, ¾-inch wood screws.

The second, inner layer of glazing can be made from one of the new lightweight films, such as Teflon (table 4–1). The film can be stretched tightly across the glazing frame assembly to make a durable (and hopefully wrinkle-free) glazing layer.

You'll need a helper when attaching the film. Place the glazing frame assembly on a flat surface. Unroll the film over the frame and cut it several inches longer and wider than the frame. The basic idea is to fold the film around the side of the frame and staple it. To do this you must first strengthen the edges of the film with duct tape or fiber-reinforced packing tape. Sand all sharp edges on the glazing frame assembly and then pull the film over the edge of the frame. One long side should be fastened first. Then staple the middle of the other long side and work toward the ends. The short ends are stapled last. When you're finished the film should be as tight as the proverbial drum.

The glazing frame assembly is installed film side down. Apply a strip of foam glazing tape to the perimeter of the glazing frame where it will rest on the collector frame. Fasten the glazing frame assembly to the collector frame with countersunk #6 x 1½-inch flathead wood screws. Make sure the screws are countersunk below the surface of the wood. Then the outer layer of glass is installed using the same method previously described. You'll need to use 1 by 2½-inch aluminum angle for the glazing caps.

Installation

Most flat plates are installed on roofs, but sometimes it's not possible or practical to install collectors up there. The alternatives are wall or freestanding installations, which are shown in figures 5–6 and 5–7. Flat plates can be installed just about anywhere so long as you keep the basic rules in mind. They should face as close to true south as is practical and have a tilt angle that's equal to the latitude. Pipe runs between collectors and storage should be kept as short as possible to minimize heat loss and pumping requirements. Of course, the collectors should always be anchored very securely so they won't be carried away be a windstorm.

There is a special consideration for wall-mounted collectors. Because the ground expands in cold weather it's important not to anchor the bottom of a collector in the ground if the top is attached to the wall of your house. In freezing weather the ground could heave, raising the collector, and something would have to give. If the top of a collector is fastened to a wall, the bottom should be fastened to the wall, too. If the bottom of the collector is rooted in the ground, the top should be, too.

The best bet for a wall installation is to use brackets made of aluminum angle, as shown in figure 5–6. These brackets can be adjusted to make corrections in the collector's altitude and azimuth angles.

Concrete footings, poured beneath the frost line, are needed for ground mounts. Three-quarter-inch galvanized pipe is inserted into the concrete while it's hardening. The collectors then can be attached to the pipe with an aluminum collar that's securely bolted to the side of the flat plate. In cold climates the bottom of the collectors should be suspended at least a foot above the ground to prevent wintertime snow blockage.

When planning a wall, ground or roof installation, keep in mind that the absorber plate must be slightly tilted to facilitate drainage. The most expensive freeze protection system will be worthless if water remains in the absorber when temperatures drop. For the same reason pipe runs should be sloped continuously from the highest point to the drain point with a drop of about ½ inch per foot of run.

Roof installations usually require a little more planning. You have to plan the physical attachment to the roof, and you also have to give some thought to the safest way to do the job. Because of its height, its slant and its lack of handholds, a roof is an inherently dangerous place. The dangers are multiplied when you start working with collectors that can weigh 100 pounds or more. You can rent scaffolding or you can build a temporary staging platform to get the job done safely.

The first step in planning is to make a sketch of your roof. Figure out not only where you'll mount the collectors but also the best place to build the scaffolding, it it's needed. Plan the job step by step.

Part of the planning involves the way in which the collectors (if there is more than one module) are plumbed to each other. The rule here is that an array of collectors should be plumbed in *parallel* rather than in *series*. In a series connection one collector would be feeding hot water to the next collector. The hotter collector would be less efficient. In a parallel arrangement the inlet temperature to all collectors is the same, and all collectors run at about the same temperature (figure 5–8).

If your roof has the proper slope for your latitude, you can mount the collectors on

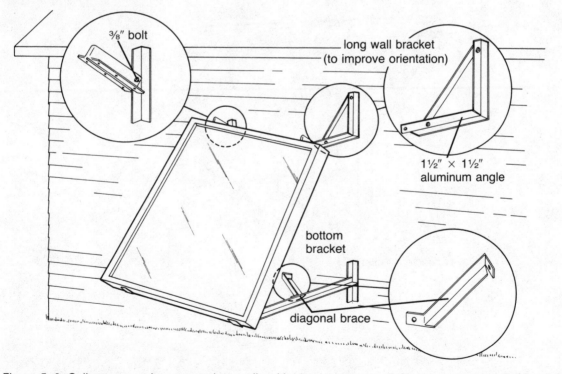

Figure 5–6: Collectors can be mounted on walls with aluminum angle. Tailor the mounting brackets so that the collectors have the proper tilt and orientation.

simple stringers. These are 2 × 2's (or 4 × 4's in areas of heavy snowfall) that are attached to the roof rafters with lag bolts. The collectors are then fixed to the stringers using aluminum angle made of ⅛-inch-thick stock (figure 5–9). The main trick to this approach is locating the roof rafters if the ends aren't exposed or if the attic is finished and the rafters are covered. If the stringers are mounted horizontally, they should be notched to allow water to pass under them. They can also be canted slightly so as not to become a dam.

Roofs that aren't pitched to the proper angle need collector tilt racks. Racks are available commercially (see Appendix 1) or you can build them yourself. If you decide to build your own support racks, the existing

pitch of your roof must be considered. Figure 3–14 explains how to find the roof pitch. The support racks can be sized by following the instructions in figure 5–10.

The collectors are strapped and screwed to the support racks, which in turn are bolted to the rafters. Lag bolts should be passed through the support racks and driven at least 2 inches into the rafters. If it's not convenient to drive the bolts into rafters, metal or wood spanners can be used. Spanners literally span two rafters, eliminating the possible hassle of locating concealed rafters. Drill through the roof, pass a suitably long bolt or length of ¼-inch threaded rod through the support rack and the roof and then fasten the bolt with nuts and washers to the spanner inside

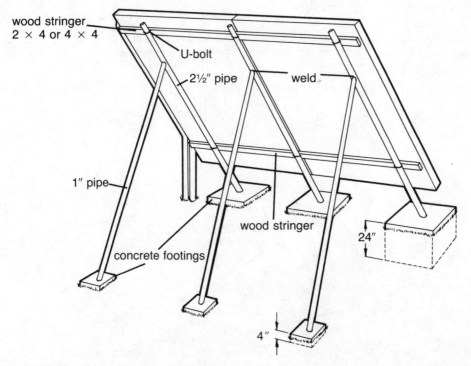

Figure 5–7: These freestanding collectors are securely braced against the wind and rain with steel pipe set in concrete footings.

Photo 5–3: Five people and three ladders can haul a heavy flat plate collector onto a roof.

the roof (photo 5–4). Afterwards caulk around all the roof penetrations to prevent leaks.

Metal struts can also be used to mount collectors on a roof (photo 5–5). Struts are sometimes easier to work with than wooden racks, especially if the flat plates must be installed in an awkward position to correct the azimuth angle.

Plumbing

One of the most important decisions you'll have to make concerns the route of the connective plumbing. The rule of thumb is to make the pipe runs as short and straight as possible. The ideal pipe run would penetrate the roof right next to the collectors and then travel straight down, inside the walls of

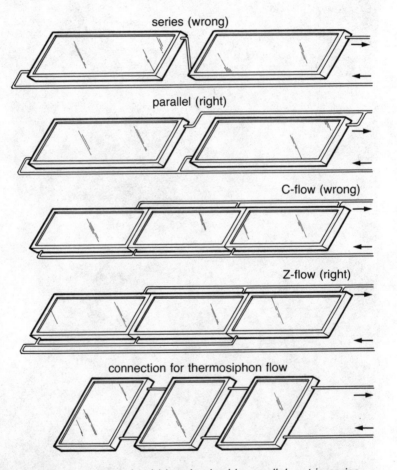

Figure 5–8: Flat plates should be plumbed in parallel, not in series. More than two flat plates should be plumbed with a "Z-flow" configuration, not a "C-flow," to encourage proper water distribution. The headers of some thermosiphon collectors are plumbed as shown in the bottom illustration.

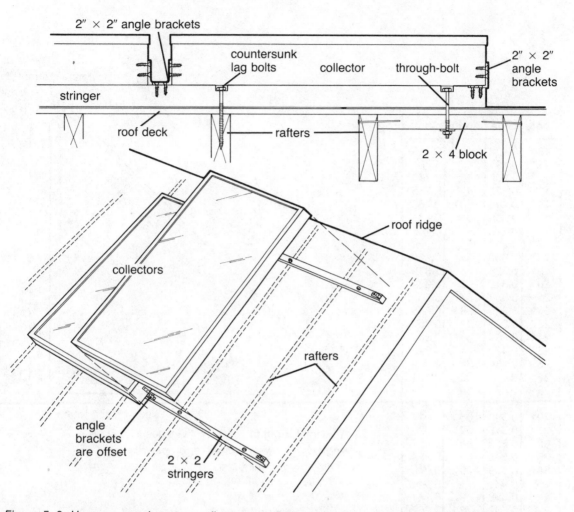

Figure 5–9: Here you see how two collectors can be installed end to end, flat on a roof. The 2 × 2 stringers are set parallel to the roof ridge and spaced a few inches less than the collector width so they can be concealed. All lag bolts are countersunk and flush with the stringers. The roof penetrations are sealed with roof patch. At lower left you see how the angle brackets are offset so both can be attached to the stringer. The upper part of the illustration shows how 2 × 4 blocks can be nailed between two rafters when no other fastening point is available.

the house, to the storage tank. But many times it's too difficult (or expensive) to rip open a wall for the tubing. You may be able to run pipes inside a closet or behind built-in cabinets. You can also run the pipes down the outside wall of the house. Each pipe should

be wrapped in the best insulation you can find. There are several types and a wide price range. The highest quality is PVC pipe filled with urethane or polyisocyanurate foam that is "prebored" to accept one or two separate runs of pipe. A more common type is a rub-

(continued on page 133)

Angle	Sine	Angle	Sine
10°	0.174	46°	0.719
11°	0.191	47°	0.731
12°	0.208	48°	0.743
13°	0.225	49°	0.755
14°	0.242	50°	0.766
15°	0.259	51°	0.777
16°	0.276	52°	0.788
17°	0.292	53°	0.799
18°	0.309	54°	0.809
19°	0.326	55°	0.819
20°	0.342	56°	0.829
21°	0.358	57°	0.839
22°	0.375	58°	0.848
23°	0.391	59°	0.857
24°	0.407	60°	0.866
25°	0.423	61°	0.875
26°	0.438	62°	0.883
27°	0.454	63°	0.891
28°	0.470	64°	0.899
29°	0.485	65°	0.906
30°	0.500	66°	0.914
31°	0.515	67°	0.921
32°	0.530	68°	0.927
33°	0.545	69°	0.934
34°	0.559	70°	0.940
35°	0.574	71°	0.946
36°	0.588	72°	0.951
37°	0.602	73°	0.956
38°	0.616	74°	0.961
39°	0.629	75°	0.966
40°	0.643	76°	0.970
41°	0.658	77°	0.974
42°	0.669	78°	0.978
43°	0.682	79°	0.982
44°	0.695	80°	0.985
45°	0.707		

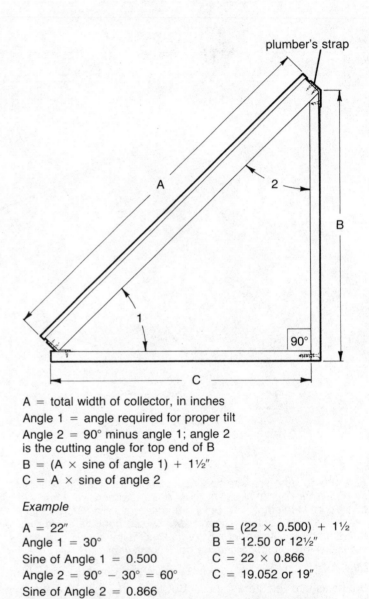

A = total width of collector, in inches

Angle 1 = angle required for proper tilt

Angle 2 = 90° minus angle 1; angle 2 is the cutting angle for top end of B

B = (A × sine of angle 1) + 1½"

C = A × sine of angle 2

Example

A = 22"
Angle 1 = 30°
Sine of Angle 1 = 0.500
Angle 2 = 90° − 30° = 60°
Sine of Angle 2 = 0.866

B = (22 × 0.500) + 1½
B = 12.50 or 12½"
C = 22 × 0.866
C = 19.052 or 19"

Figure 5–10: This calculation shows how to figure out the dimensions for making a wooden collector rack that's customized to your collector dimensions and to the tilt you need. If, for example, your roof has a 20° pitch and you need a 42° tilt, build the rack so that Angle 1 equals 22°. Plumber's strap and screws are used to fasten the collectors to the rack.

Photo 5–4: When it's not convenient to bolt a collector to the rafters, a spanner board such as this is the next best thing.

Photo 5–5: Metal struts are often used as collector racks.

Photo 5–6: Sometimes it's easier to run the pipe down the outside wall of the house. In this installation the pipes were insulated and encased in 4-inch-diameter PVC drainpipe before they were installed.

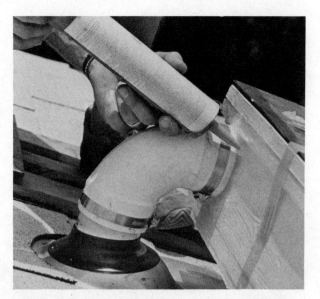

Photo 5–7: Roof penetrations should be weather-tight. Here the collector outlet pipe, hidden behind heat tape, insulation and a PVC pipe elbow, disappears into the roof through a neoprene boot. All cracks and joints should be sealed with silicone caulk.

bery black foam that is a good insulator, but isn't made to withstand exposure to the elements. When using this type, the pipe runs should be encased in PVC pipe, which can also be used (in active systems) to harbor sensor wires.

Rather than running the pipes down to a water heater or storage tank in the basement, you also have the option of reinstalling the tank in the attic or on the second floor, if you can find room for it. This approach may have other benefits. For example, the closer an active system's collectors are to the storage tank, the less pumping power is needed to move the collection fluid.

One of the trickiest parts of the installation is the roof (or wall) penetration, where the plumbing enters the house. If the penetration point isn't weathertight, you'll live to regret it. There are several good penetration kits on the market (see Appendix 1). A special neoprene boot is commonly used to seal around the pipes as they pass through the roof (photo 5–7). The shingles are rearranged around the boot so that water is carried off, and everything is covered with a healthy amount of roof "goop" (butyl roof patch). It's a good idea to have one large penetration for both hot and cold pipes, rather than two smaller penetrations. Wall penetrations should use a metal or plastic sleeve around the pipes where they pass through the foundation or the house wall (figure 5–11).

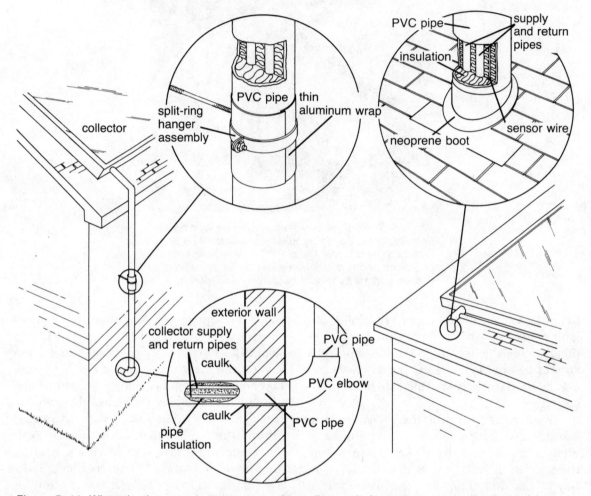

Figure 5–11: When the time comes to penetrate the wall or roof of your house, take the time to do a good job. Standard neoprene boots are available at hardware building supply stores. Use plenty of caulk.

6

PASSIVE AND ACTIVE FLAT PLATE SYSTEMS

Now that you have your collectors, you're ready to install the system that will put them to work. This chapter covers four flat plate systems that were introduced in chapter 2: a passive thermosiphon system, an active draindown system, an active drainback system and an active antifreeze system. Also discussed is the recently developed *phase change system*, which uses a fluid that changes from a liquid to a vapor and back to a liquid as it flows between collector and storage. An even more recent development makes a new power source available for pumps: the sun. *Photovoltaic*—or *solar cell*-powered—pumps take the electricity bill out of running an active system.

Thermosiphon systems have no moving parts. They work because warm water is lighter than cooler water. In this system the sun is the pump.

The open-loop draindown system uses a pump to transport water from the storage tank to the collectors and back. It incorporates a thermostatic controller, temperature sensors and a special valve that works automatically to drain the collectors in the event of freezing weather and other hazardous conditions. When the sensors detect a freeze, a message is sent to the valve, which closes to cut water pressure and allow the collectors to "drain down."

The drainback system uses one or two pumps. It has two plumbing loops. Unpressurized water is circulated between the collectors and a small tank in the solar loop, and the resulting heat gain is transferred to the second loop which contains the potable water by means of a heat exchanger. This system also has a controller and sensors, but no automatic valves. Whenever the pump stops, the water in the solar loop drains from the collectors into a holding tank.

In antifreeze systems the solar loop is filled with antifreeze, so there is no need to drain the collectors when the outside temperature approaches freezing. As in the drainback system, a heat exchanger is used to transfer heat from the antifreeze to the domestic water.

The phase change system is a somewhat new development in solar water heating. It also uses a controller, a pump and a heat exchanger, but instead of employing an antifreeze fluid, a phase change system uses Freon. Liquid Freon is pumped to the collectors, where it absorbs heat and turns to a gas, "changing its phase." The gaseous Freon then returns to a heat exchanger immersed in a domestic water tank, where it gives up its heat. The Freon then condenses back to a liquid, again "changing its phase," and is circulated back to the collectors. Other phase

Photo 6–1: Collector installations tend to be pretty site-specific, but as a rule there should be at least two people on hand and plenty of ladders to get people up to the work site and help them stay there.

change systems use no pump and run instead like a water-filled thermosiphon system.

D-I-Y or H-I-O (Hire It Out)?

A homeowner with some basic carpentry, plumbing and electrical skills can install the first three, water-based systems, and possibly even an antifreeze system. But the phase change system shouldn't be attempted by a do-it-yourselfer. In this system the collectors and plumbing incorporate special components, and the loop must be filled with liquid Freon, something only a refrigeration tradesman can do properly. The phase change system is described here so you can get a good idea of how it works. You'll be armed with enough information so that, if this system is what you want, you can hire an installer who has experience with it. The antifreeze system isn't nearly as tricky to install, and if you're looking for someone to sell, install and guarantee a system, chances are the system will be an antifreeze type. Of course, solar installers also work with the water-filled systems, though they aren't yet quite as popular in the trade.

Thermosiphon, draindown and drainback systems have long been successfully installed by do-it-yourselfers. They've used home-built and store-bought collectors, and they've pieced together the special solar components and standard plumbing and electrical hardware to come up with efficient, reliable systems. It wasn't always easy. In the early 70s there were no "special solar components" and some degree of jury-rigging was called for. But nowadays you can buy completely assembled components—from commercially made collectors to factory-assembled control modules—and install them yourself. You've also seen how you can build your own collectors, and in this chapter you'll see that you can assemble your own control module (controller, pump, sensors and other parts) from off-the-shelf solar components.

It's certainly not assumed that you've made a decision on a system. The following sections will help you do that with their descriptions of the systems, their installation and operation requirements and their suitability for different climates. To give you more insight into the decision-making process and the vagaries of site-specific installations, there are also a couple of real-world case studies of installations that were done by Rodale's Research and Development Department. "Site-specific" is a key word. Every retrofit installation presents unique problems of getting collectors, plumbing, tanks and wiring installed neatly and unobtrusively, and with a minimum of upset to the house and its occupants. But there are "generic" installation procedures, such as wall and roof penetrations and getting pipes inside an existing wall, that can be explained to make your job easier and to help you and have a better result.

Thermosiphon

Minor Street in Emmaus, Pennsylvania, is lined by tall trees and many old prewar houses. One of these houses was recently converted into the editorial offices of Rodale's *New Shelter* magazine. *New Shelter*, as informed do-it-yourselfers know, is a publication devoted to energy-efficient house design and to sensible home improvements. Solar heating has always been dear to the hearts of its editors, and so the decision was made to fit the house with a solar water heater.

The design of the house presented some problems. Only a small portion of the roof faced true south, with most of the roof facing either east or west. And much of the south-facing roof was shaded by the house next door, with only a small portion receiving almost a full day's worth of sunshine. The south yard, incidentally, was even more hopeless from a solar point of view. Tall deciduous and evergreen trees, as well as the house next door, made it especially shady and un-solar.

The house also had some good characteristics. The attic was unused, and part of it was next to the unshaded roof section. It looked as though a thermosiphon system could be installed because this type of system requires that the storage tank be higher than the collectors. (More precisely, the bottom of the tank must be at least 18 inches above the top of the collectors.) This allows the water to circulate by natural convection, without pumps. The tank not only must be installed

above the collectors, it also must be relatively near the collectors, with 30 feet, at most, to ensure a good flow rate. The greater the separation, the more pipe friction plays a role in slowing the flow of water and thereby reducing the system's efficiency.

At the *New Shelter* house, it turned out that the roof pitch ran low enough, and the attic rafters high enough, to permit installing the tank above the collectors, with a separation of only 10 feet. The fact that the attic was unheated wasn't a problem because thermosiphon systems with exposed collectors must be drained before freezing weather arrives. If a year-round system, such as a drain-down system, had been chosen, the pipe runs through the attic would have had to be pro-

Photo 6–2: This home had attic space above parts of the roof, which made it possible to install a thermosiphon system. The storage tank was braced horizontally in the attic.

tected with heat tape, or the attic heated. Pipe insulation alone would not be enough.

However, it is possible, to digress a little, to design a thermosiphon system that can run all year with no risk of freeze-up problems. The key is to put the collector, the tank and all the plumbing indoors. That sounds pretty normal for the tank and pipes, but an indoor collector? If you've ever stood on the inside of a sun-washed window, you have been an indoor collector, picking up an extra blast of heat over what the radiators deliver. You've probably never frozen indoors, either. A thermosiphon collector will enjoy the same benefits if you put it behind a window or a skylight, or in an attached solar greenhouse, if you have one. The absorber plate shouldn't be encased in the airtight box described in chapter 5, because that would cut it off from the protective heat of the living space. The box could be unglazed, or the absorber plate could be unboxed, and either would have to be held back a foot or so, from the house glass, especially the box, so it could be surrounded by warm house air. The disadvantage of this is that an open collector won't heat up as much as the hermetic model. But it lets you keep a thermosiphon system in operation all the time, and the moving parts are water molecules.

This approach wasn't taken at the *New Shelter* house because it wasn't needed. The collectors were installed outdoors because solar-heated water wasn't needed when the weather got cold. The renovation of the house included a new coal-fired heating system that, during the heating season, produces great amounts of hot water for the staff as well as for the radiators.

Thanks to technological progress, there are ways to keep exposed thermosiphon collectors in operation year-round with no fear of freezing. A thermostatically controlled freeze protection valve, made by the Eaton Corporation of Carol Stream, Illinois (see Appendix 1), is designed to open when the collector water temperature drops below 45°F, thereby "bleeding" water from the collectors. The pressurized collectors are immediately refilled with water from the cold water supply line, which is plumbed between the storage tank and the collector inlet. The replacement water doesn't come from the storage tank, but from the house cold water line. Even in winter, this water is warm enough to prevent freeze-ups. The valve uses no electricity, making it ideal for passive systems.

The Installation

To start, a drawing of the roof and the attic was made to help plan the placement of the collector and the tank (figure 6–1). As it turned out, the collectors had to be mounted so that their plumbing connections could enter the attic directly without coming in through a downstairs room. This left little room for installing the tank. In fact, there wasn't enough room to install it vertically, so it was installed horizontally. This presented a slight problem, because an important factor in the proper operation of a thermosiphon system is having the cold and hot water literally be separate, or *stratified*, inside the tank, so that the coldest water is always delivered from the bottom of the tank to the collectors. It's always best to install a thermosiphon storage tank vertically, for maximum stratification. Because the tank in the *New Shelter* house had to be installed horizontally, special precautions were taken to make sure the hot and cold water inside the tank would separate adequately, but more on that shortly.

Once the collector location was established, the pitch of the roof was measured.

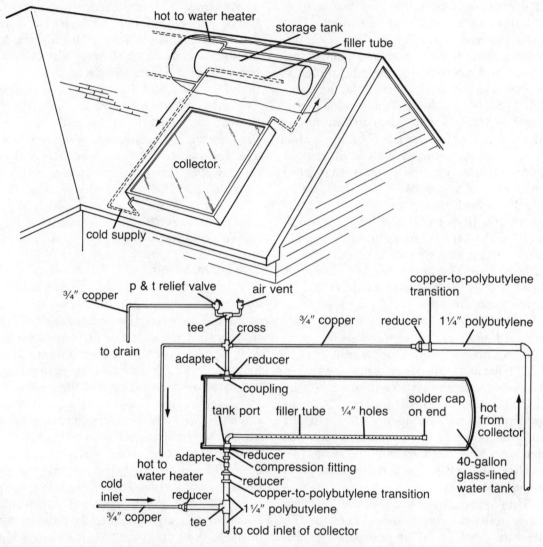

Figure 6–1: A thermosiphon system is a model of simplicity in action. Here the tank is braced horizontally against the rafters. Whenever the tank is installed horizontally a filler tube is needed to maintain temperature stratification. The hot water outlet pipe leads to a conventional water heater in the basement.

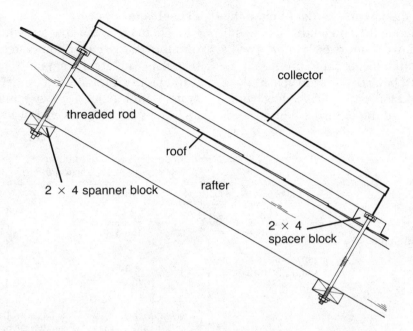

Figure 6–2: In this installation 2 x 4 spanner blocks were used to hold down the collectors when the rafters proved to be inconveniently spaced.

You'll recall that the angle of the collectors should be within 5 degrees of the latitude, which turned out to be the case at the *New Shelter* house, allowing the collectors to be mounted flush with the roof.

The actual installation method was similar to the one described in chapter 5. Instead of using long stringers, short, pressure-treated 2 × 4 blocks were fastened to the corners of the collector. Normally the blocks would be lag-bolted through the roof to the rafters, but this proved to be impossible because of their irregular spacing. The solution, however, was simple. Holes were drilled through the blocks and through the roof *between* the rafters. Then four 2 × 4 spanner boards were cut to be longer than the rafter spacing. Then a long, threaded rod was passed through each block and through the roof. The threaded rods were long enough to protrude beyond the rafters and through the spanner boards. Nuts and washers were then all that were needed to lock the collectors in place (figure 6–1). Outside on the roof, healthy dollops of butyl roof patch were applied around the blocks to prevent leakage into the attic.

Figure 6–3 shows the standard schematic for a thermosiphon system. The cold water inlet to the collector is at the bottom, and the hot water outlet is at the top. The risers on the absorber plate run slightly uphill toward the outlet. This helps to maintain an adequate flow rate and is an important design detail. In this installation the risers were slanted to rise at least ¼ inch for every 2 feet of run. This slight canting also makes it easier to drain the collector. With the risers running nearly horizontal, the perpendicular headers naturally run straight up the roof pitch. If the situation demanded it, you could also install your collector with the risers running with the roof pitch and the headers running across it. It would still be a good idea to give a slight cant to the collector so that the headers would run uphill toward the outlet.

Plumbing the System

There are some special requirements for plumbing a thermosiphon system. Unlike active systems that rely on pumps to force water through the plumbing, sun-driven passive systems don't develop very much "push," and steps must be taken not to impede the

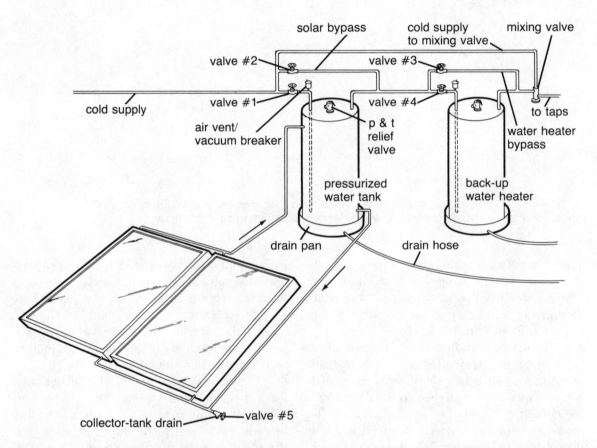

Figure 6–3: In a standard thermosiphon system the tops of the collectors are at least 18 inches below the bottom of the storage tank, which is plumbed to the existing water heater. Valves are used to by-pass the storage tank and the water heater, or to use them together. In the winter the thermosiphon loop is drained by closing valve #1 and opening valve #5. In the summer it's possible to shut down the conventional heater by opening valve #3 and closing valve #4. The mixing valve is needed in case the thermosiphon-heated water gets too hot. Most months of the year the storage tank and the conventional heater are used together by opening valves #1 and #4 and closing valves #2, #3 and #5. The thermosiphon loop then acts as a preheater. The check valve prevents reverse thermosiphoning.

already delicate flow. The pipes in the solar loop must be extra large to minimize friction effects and, for the same reason, bends and curves in the pipe runs must be kept to an absolute minimum.

Pipe runs of up to 20 feet between the collector and the storage tank should be made with 1-inch-i.d. tubing. Pipe runs of up to 30 feet in length require 1-¼-inch-i.d. tubing. For longer runs there will be little benefit to using a larger pipe size, though 1½-inch-i.d. pipe has been used. The point is to minimize the collector-to-storage distance.

At the *New Shelter* house 1¼-inch-i.d. copper tubing was used outside and a couple of feet into the attic. This exposed metal tubing was protected by heat tape and 3-inch-diameter pipe insulation. The heat tape is controlled by a little thermostat that is placed outside, close to the inlet and outlet (two separate heat tapes were used). The other end of the tape is plugged into a 120-VAC outlet.

To keep costs down, copper pipe was used only outside and through the roof penetrations. Inside the attic 1¼-inch-i.d. *polybutylene* tubing was joined to the copper pipe with special adapters. The polybutylene tubing was joined to very short sections of copper tubing at the storage tank.

Polybutylene plumbing is a recent innovation for residential water systems. It is code-approved for both cold and hot water service. It is also somewhat easier to work with than copper, and the materials for a polybutylene system will be considerably cheaper than those for the same system done with copper. One of its nicest characteristics is that if water happens to freeze inside a pipe section, the polybutylene won't burst; it'll just stretch and then return to its original size when the ice melts.

The strategy for minimizing sharp bends in the pipe runs involved using 45-degree elbows instead of right-angle elbows. Even when the pipe had to make a right-angle turn, two 45-degree elbows were used instead of a single 90-degree elbow, all toward the goal of minimizing pipe friction.

Photo 6–3: This is how the tank was installed horizontally in the attic of the house shown in photo 6–2. The rack was made from 2 x 4's, 2 x 6's and 2 x 8's.

The final outside work involved sealing the roof penetrations made by the inlet and outlet pipes. Three-inch-diameter neoprene roof boots were previously installed around the insulated pipe and then integrated with the shingles. For double protection the penetrations were also sealed with butyl roof patch.

The solar storage tank in a thermosiphon system essentially serves as a preheater: When a tap is opened, solar-heated water flows from the storage tank to the conventional water heater (which is often some distance away). Depending on how hot this water is, the water heater will either be kept off, or it will have to fire up a little to boost the temperature. At the *New Shelter* house a 40-gallon tank was installed. Since the collector had an area of about 32 square feet, a 1.25 to 1 (square feet per gallon) ratio was about right for this region.

The tank was braced against a wall in the attic as shown in photo 6–3. Since the filled tank weighs about 400 pounds, care was taken to make the support brackets very strong. You'll recall that the preferred tank position is vertical, which allows the tank to stand on its own without the need for any bracing. Whether you're placing it horizontally or vertically, make sure to leave enough room at the top of the tank for insulation and plumbing fittings.

If you must install the tank horizontally, you'll have to take special measures to preserve temperature stratification inside the tank. In a vertically installed tank, hot water stays at the top where it is drawn off, while cooler water remains at the bottom. But when the tank is lying on its side, it's more difficult to prevent the hot from mixing with the cold, especially when cold water enters to replace a hot water draw. To maintain proper temperature stratification you can make a *filler tube* (figure 6–1). This is simply a cold water inlet pipe that directs cold water to the bottom of the tank. It is made from a piece of ¾-inch copper tubing that's cut to fit within the length of the tank. One end of the filler tube is passed through the inlet port and attached to the cold water inlet pipe with a ferrule-type compression fitting that has been drilled out to permit the filler tube to pass completely through. The other end of the tube is closed with a soldered cap. Drill a dozen or so ¼-inch holes along one side of the filler. These holes allow cold water to enter the tank with a minimum amount of turbulence. When installing the filler tube, make sure the holes face downward. A vertically installed tank doesn't need a filler tube. The cold water inlet pipe is plumbed directly to the inlet port, so long as the port is near the bottom.

The pressure-and-temperature relief valve at the top of the tank is standard for any solar tank, as is the air vent.

The hot outlet from the tank leads to the existing water heater, which was located in the basement at the *New Shelter* house. At the time, the whole house was being remodeled, and many partition walls had been stripped of their old plaster and lath, awaiting new Sheetrock. With the open walls it was easy to drill holes between floors and run the pipes inside the stud walls. This is something to remember if you've got both remodeling and a solar water heater in your future. If you've got only the latter event planned, you can still conceal pipes in a stud wall by using long drill bit extensions to get through top and bottom plates and any fireblocks there might be. Other strategies include running pipes through closets and through the backs of built-in cabinets.

The valving arrangement shown in figure 6–3 allows the solar water heater to work alone as the sole heat source or as a preheater in concert with the existing water heater. It also allows the collectors to be drained with the onset of cold weather. The cold water supply is plumbed between the solar tank and the collector. To fill the system, open valve #1 and close valve #5. To drain the system, close #1 and open #5. In operation, the sun heats the water in the collector, which rises and flows into the top half of the solar tank. Cooler water at the bottom of the tank simultaneously circulates back to the collector. It's as simple as that.

The system described here has been operating successfully for several years. The tank and collectors are drained and shut down well before the first frost of the year, and refilled once the danger of freezing has passed.

Photo 6–4: Sometimes it's difficult or impossible to install a thermosiphon system on the roof of a house. For such situations Environment|One of Schenectady, New York (see Appendix 1), makes this ground-mounted thermosiphon system. The tank lies on its side beneath the shingles at the top of this wedge-shaped unit.

The heat tape acts as insurance against an early or late frost.

Draindown

At another one of Rodale Press's small house-become-offices in Emmaus there was no way that a thermosiphon system could be installed. The south-facing lawn was completely shaded by a neighboring house, and the roof was flat, with no place to put a storage tank above the collectors. But that was OK. The roof received plenty of sunshine and, because it was flat, collectors could be precisely positioned to face true south. The solar potential was there, and an active system was needed.

Once it was decided to install an active system, the next step was to select a method of freeze protection: either draindown, drainback, or antifreeze. Draindown freeze protection was selected because the Rodale Research and Development people wanted to study it as part of their work with all types of solar DHW systems. For the homeowner the draindown approach can be less expensive to install than a drainback or antifreeze system, although, after the thermosiphon system, the drainback is probably the simplest of flat plate systems.

Photo 6–5: A draindown valve is electrically powered by the differential controller. When there's no solar gain and temperatures drop toward freezing, it drains the collectors.

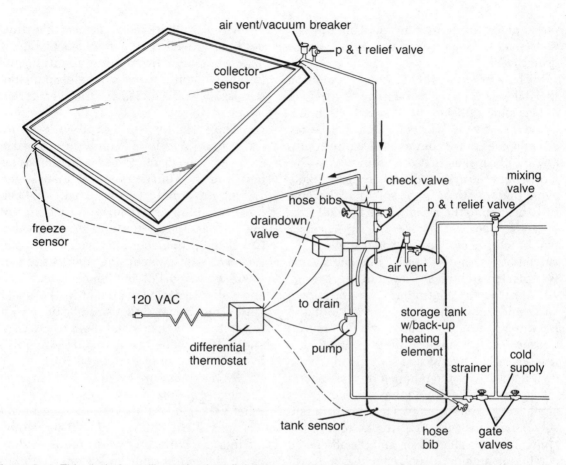

Figure 6–4: This draindown illustration is split into two sections. The components above the break in the pipe lines are installed outdoors, while everything below the break should be installed in an indoor space that doesn't experience freezing temperatures. The differential thermostat runs the system: It starts the pump when the collectors are hot enough and opens the draindown valve, draining the collectors, when the outside air temperature approaches freezing.

Like any active system, draindown systems work by pumping water between a storage tank (which can be an existing water heater) and the collectors.

The heart of the freeze protection system is an electrically powered draindown valve. When the pump is operating, the draindown valve is open, and the solar loop is filled with pressurized water (open-loop system.) When the pump stops running because there isn't enough solar gain, the valve stays open, and the collectors stay filled. When an outside sensor detects a temperature drop toward 32°F, it signals the controller to cut power to the valve, which closes and cuts off water pressure to the collectors. When the valve closes, a little drain port in the valve is opened, and all the water in the collectors and the exposed plumbing drains down and out through the valve, through a tube and into a house drain.

There are two other modes of operation that take care of the possibility of a power outage and a collector overheat situation. Because the valve is "normally closed" (the position it goes to without power), a power outage simply means that the draindown valve closes, and the system drains. In the unlikely event that the water in the solar tank became fully heated, to the point where the collector began to overheat (above 180°F), a sensor would tell the system controller to close the valve (i.e., cut power to it). Thus, the system is protected in no less than four ways.

What is the controller? The draindown valve and the pump both are wired to it and it's more commonly known as a *differential thermostat*. It is an electronic device that is used in all active systems. Basically it's a switch that turns the pump and valve on and off. It knows when to do that by reading the signals it receives from two temperature sensors. One sensor is put at the bottom of the conventional water heater, and the other sensor is put outside, usually on the collector outlet (figure 6–4). When the temperature of the collectors is higher than that of the water in the bottom of the water heater, the differential thermostat turns on the pump, and, in the case of this system, it powers the draindown valve to an open position. The thermostat also controls the freeze, power outage and overheat modes just described.

The Installation

As usual, the collectors were installed first. They were hauled to the roof by the double-ladder method shown in photo 6–6, then mounted to the roof using homemade support racks like the ones described in chapter 5. It turned out there was a slight problem with these racks because they didn't hold the collectors far enough above the roof to prevent snow from piling up on the glazing. They should have been blocked up so as to raise the collectors above the average level of snowfall.

The system was plumbed with ½-inch-i.d. copper. A rule of thumb on pipe sizing is that if the distance between the storage tank and the collectors is less than 40 feet, ½-inch-i.d. tubing can be used. Pipe runs longer than 40 feet require ¾-inch-i.d. tubing. Copper tubing is the most commonly used; CPVC tubing (which is approved for hot water systems) is less expensive than copper and it will work, too. Polybutylene tubing, fittings and valves are also available in these standard sizes. Because they are plastic and can sag more easily than metal tubing, both CPVC and polybutylene tubing require support at least every 2 feet of horizontal run.

At the highest point in the system, at the collector outlet, a pressure-and-temperature relief valve and an air vent/vacuum breaker were installed. The vacuum breaker admits air into the collectors when the draindown valve closes to drain the system. Without a vacuum breaker the collectors probably wouldn't drain. Heat tape was then wrapped around the vacuum breaker to eliminate the possibility that it might freeze up and not open for draining in the winter. The relief valve (p&t valve in the schematics) is standard protection for all collectors. If for some reason the collector became overpressurized (over 75 psi) or too hot (over 190° to 200°F), the relief valve would pop open and pressurized water would flow through the collector until it cooled down and/or the overpressure was relieved. You'll find the same kind of valve on your water heater, serving the same purpose. The only exception is with the antifreeze system, which uses a pressure-

Photo 6–6: This is an installation method using three people, two ladders and a rope. Everyone should have a hard hat.

relief valve, not a pressure-and-temperature valve.

In this installation it was decided to run the connective plumbing down the outside wall of the house (photo 5–6), rather than through the roof and down through two floors to the basement. The plumbing finally enters the house close to ground level, penetrating a low roof over a stairway leading to the basement. This method saved the installers much of the trouble of routing the plumbing inside, and concealing it in interior walls.

It is important to note that not only the collectors, but also the exposed plumbing should have a continuous downhill slope, no ups and downs, or the collectors may not drain properly. The recommended pipe slope is at least ½ inch for every foot of run.

Pumps

As the schematic in figure 6–4 shows, the pump is located between the tank and the draindown valve so that it won't impede drainage. There are dozens of solar pumps now on the market with a wide variety of power ratings, flow characteristics, even with multispeed controls. It is important that a pump not be too overpowered or underpowered for a given system and there are two primary factors to consider when sizing and choosing a pump. One is the flow rate developed by the pump (measured in gallons per minute, or gpm). The second is the force or resistance the pump must push against to circulate water between the storage tank and the collectors (this is called *head pressure*). Head pressure is the sum of two distinct components: *friction head* and *gravity head*. The former refers to the resistance to flow caused by the frictional effects of water rubbing against pipe walls and passing through valves and other fittings. Gravity head refers to the vertical height to which the fluid must be

lifted by the pump and (for our purposes) is a factor that is present only in drainback systems. In all other systems gravity head is zero because fluid flowing down from the collector counterbalances the fluid going up. The drainback system is the only one in which the solar loop isn't filled all the time (the draindown system fills itself from water line pressure). Head pressure is usually expressed in terms of either "feet of water" or pounds per square inch (psi). To convert to psi, divide feet of water by 2.31, and to convert to feet of water, multiply psi times 2.31. It's pretty easy to quantify both flow rate and head pressure in terms of your own system so you'll be able to select the right pump.

The required flow rate is determined by the area of the collector. As a rule of thumb, there should be a flow of 0.03 gpm for every square foot of collector. Thus, a system with 100 square feet of collector area needs a pump that will produce a flow rate of 3 gpm (100 × 0.03 = 3). But this isn't all. There are other factors that come into play in determining the head pressure for any system, drainback or otherwise. The following formula is used for all systems *except* the drainback.

head pressure (ft.) = friction head (ft.)

$$= \left[\frac{1}{50} \times \begin{array}{c} \text{total} \\ \text{length} \\ \text{of} \\ \text{pipe} \\ \text{(ft.)} \end{array} \times \begin{array}{c} \text{flow} \\ \text{rate} \\ \text{(gpm)} \end{array} \right] + 4$$

Since there is no gravity head factor for these systems, head pressure essentially equals the calculated friction head. You know what your required flow rate is, so all that remains is to add up the total length of your pipe runs. This length includes all vertical and horizontal distances in both directions of the pumped loop. The "plus 4" at the end of the

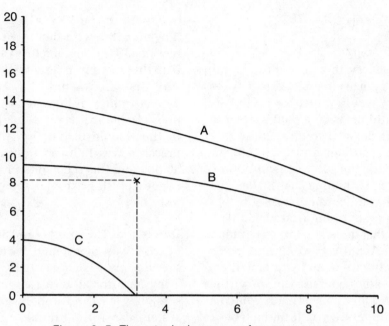

Figure 6–5: Three typical pump performance curves.

equation accounts for the friction effects of valves, fittings and the collector itself. If, for example, the total pipe run for a system were 70 feet, the equation would be:

$$\begin{array}{c} \text{head} \\ \text{pressure} \\ \text{(ft.)} \end{array} = \left(\dfrac{1}{50} \times 70 \times 3 \right)$$

$$+\ 4\ =\ 8.2$$

With the flow rate and head pressure determined for this system, you could go ahead and specify a pump using just these two numbers: a pump that can deliver 3 gpm against a head pressure of 8.2 feet of water. Typical pumps that are suitable for this group of systems have electrical ratings of about $\frac{1}{50}$ to $\frac{1}{10}$ hp and draw about 40 to 160 watts of electrical power.

If you are planning to install and equip a drainback system, use the following formula to determine head pressure:

$$\begin{array}{c}\text{head}\\\text{pressure}\end{array} = \begin{array}{c}\text{friction}\\\text{head}\end{array} + \begin{array}{c}\text{gravity}\\\text{head}\end{array}$$

$$\begin{array}{c}\text{head}\\\text{pressure}\end{array} = \left[\dfrac{1}{50} \times \begin{array}{c}\text{total length}\\\text{of pipe}\\\text{run (ft.)}\end{array} \times \begin{array}{c}\text{flow}\\\text{rate}\\\text{(gpm)}\end{array} \right]$$

$$+ \left[\begin{array}{c}\text{vertical distance between pump}\\\text{and collectors (ft.)}\end{array} \right] + 4$$

Gravity head is found by measuring the feet of vertical distance between the pump and the top of the collectors, assuming that the collectors are the highest point in the system. Don't include any measurement of horizontal distance. The friction head is calculated in the same way as in the previous equation. Suppose that a gravity head of 20 feet is measured for a system with the same friction head as in the previous equation. Then the equation becomes:

$$\text{head pressure (ft.)} = \left(\frac{1}{50} \times 70 \times 3\right) + (20) + 4 = 28.2$$

For this drainback system the required pump must deliver 3 gpm against 28.2 feet of water, which typically would indicate a $\frac{1}{20}$ to $\frac{1}{4}$ hp pump that would draw 80 to 280 watts. Naturally you will do well to choose the most efficient pump you can find. A pump that does the job with the least consumption of electrical power is the one that makes your solar system that much more cost-effective.

When you're shopping around you'll undoubtedly find that there are dozens of pumps to choose from. What you won't find is very much uniformity in manufacturers' literature. They will all of course extol the virtues of their own product, but how do you compare? In most literature you'll find a table or a chart of a pump's actual performance, or what flow rates it delivers against different head pressures. In figure 6–5 we copied some of the curves of popular solar pumps to show what they're like. The axes of the graph use flow rate and head pressure, so you can locate your point in the quadrant using the numbers you've come up with. Say, for example, that the pump specs from the first equation are plotted. The "X" in the illustration shows where 3 gpm and 8.2 feet of head pressure cross. Pump B is the obvious choice. You could buy pump A, but you'd be needlessly using 10 more watts to get the job done.

Draindown Valves

You don't have a great deal of selection in draindown valves because at this writing there are just a few manufacturers that have them in production (see Appendix 1). They are all similar in the way they work, and they typically accept $\frac{1}{2}$-inch and/or $\frac{3}{4}$-inch pipes. They operate on standard 120-VAC house current (one manufacturer, Sunspool, also has

one that runs on 12 VDC, direct current) and they all come with their own installation directions. The manufacturer's literature also lists the maximum flow rate and water pressure the valve is built for, although testing by researchers at Rodale has shown that the pressure claims have been a little excessive. In the Rodale installation the house water pressure was reduced from 80 psi to about 50 psi, and this seemed to improve the opening and closing action of the valves that were tested.

Differential Thermostats and Sensors

As was mentioned earlier, differential thermostats all do essentially the same thing: They monitor at least two temperature sensors, they start and stop the pump and in this system cause the draindown valve to open and close. The thermostat should, of course, run on the same voltage as the pump and the draindown valve (120 VAC is by far the most common). The manufacturer will specify the maximum current (in amps) that its unit can handle to run the pump and, in this system, the draindown valve. For example, the combined load of the pump and the valve in this system was a maximum of 1.3 amps, and the thermostat that was used could handle a load of 1.5 amps.

The temperature sensors normally come with the differential unit. Some are just tiny plastic-coated bulbs that you can tape directly to the water tank and the collector outlet, while others are made to be screwed into plumbing tees to read the fluid temperature directly. It's very important that the sensors have absolute contact with the surface that they're sensing. With pipes, sensors should be taped very tightly. Epoxy can be used to bond a sensor to the wall of a tank.

You can order the differential thermostat, sensors, draindown valve and pump from a local solar supplier or you can get them through one of the sources listed in Appendix 1. When buying locally, double-check with the supplier to make sure that the components you have in mind are compatible. If you still have doubts you can always drop a line to the manufacturer.

The Tank

To make this installation less expensive, the existing water heater was converted to be both a solar storage tank and a back-up water heater. The lower heating element was disconnected to maintain better stratification inside the tank. This simply meant disconnecting the wires leading to the element,

making sure the power source was disconnected beforehand. The drain valve at the bottom of the heater was replumbed to be the cold supply to the tank and the collectors as well as a drain. What was formerly the supply at the top of the tank became the hot water return from the collectors (figure 6–7). A check valve was added between the heater and the collectors to prevent water from flowing backward. The remaining components, plumbing connections and valves were installed as shown in figure 6–4. The draindown valve's drain outlet was connected by a small plastic tube to a nearby sewer drain.

After the various components were installed and all the connections soldered, the

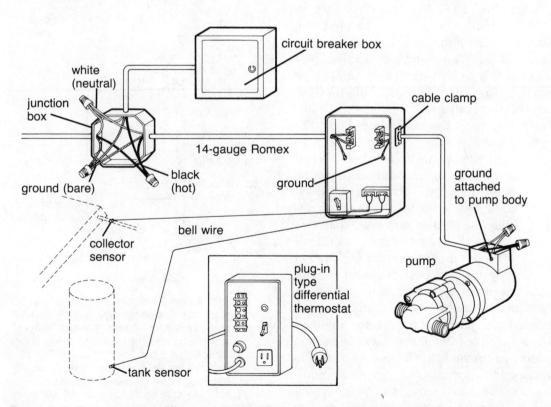

Figure 6–6: The pump and differential thermostat are wired with standard code-approved procedures.

system was filled. The electrical components were activated, and the plumbing was checked for leaks. Once it was certain there were no leaks, the pipe runs were covered with ⅝-inch-i.d. by ¾-inch (wall thickness) closed-cell neoprene foam pipe insulation. Outside the house, this insulation was covered, for protection and better looks, with 4-inch PVC pipe. This made the pipe run look more like a gutter downspout, much *less* to look at. Heat tape was wrapped around several inches of exposed tubing near the collector inlet and outlet ports as well as around the air vent/ vacuum breaker. The wires for the heat tape and the sensors were all run through the PVC pipe.

Builder Beware

If you think about it, a draindown system may not be totally invulnerable to a freezing "episode." What if the valve jammed and was unable to close? You might well ask. The best answer is that these valves have undergone years of testing and refinement, and they *are* ready for your system. Along with their built-in safety features, they can claim high reliability. There's not much more that can be done about the vacuum breaker beyond wrapping it with heat tape and making sure, from time to time, that the heat tape works.

You might want to consider buying a differential thermostat that uses more than two sensors. Most differential controls monitor a sensor at the bottom of the water tank and at the collector outlet. A possible problem with this arrangement is that the collector sensor might not be able to sense quickly enough when a lower part of the collector might be cooling to the point where ice was beginning to form. To prevent this from occurring you can use a differential control with an additional "snap freeze" sensor that monitors absorber plate temperature. If the sensor detects

a cold snap with no solar gain, it will cause the collector to drain. Air temperature is not a reliable indicator of absorber temperature because under some conditions the rate of

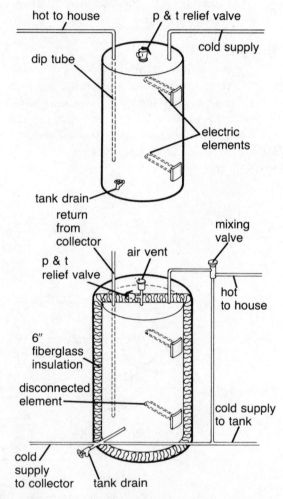

Figure 6–7: A conventional water heater can be transformed into a solar storage tank by converting the tank drain into a combined cold water supply to the tank and the collector. The bottom heating element (if it's an electric heater) is disconnected to promote good temperature stratification. A mixing valve is added to prevent scalds, and the tank and pipe runs are all well insulated.

Photo 6–7: The finishing touches are applied to this draindown system. The differential controller was mounted on the wall to the right of the electric meter. The draindown valve is near the man's forehead, and the pump is near his knees. The pump is secured against a wall or other solid object to minimize pipe rattle. The vacuum breaker plumbed at the top of the water heater allows faster tank draining.

radiant heat loss from the absorber can cause it to become colder than the outside air. The collector could freeze before the ambient temperature went below 32°F. H. I. Square, Inc., of West Palm Beach, Florida (see Appendix 1), makes a differential control (called the Fixflo Control with Frost Override, Model #H–15–6–C) that can accommodate up to four sensors. The fourth sensor detects unusual nighttime frost conditions (figure 6–8).

Drainback

Drainback systems have many of the advantages of draindown systems, but don't have some of the disadvantages. The collectors in a draindown system could freeze if the draindown valve jammed and couldn't close, or if the vacuum breaker didn't open. Drainback systems aren't bothered by these problems because they don't have electric valves, or even air vents. The drainback motto is this: Whenever the pump is off, the collectors are empty.

How's that? A typical drainback system has two plumbing loops. One is the *solar loop*, which connects the collectors to a holding

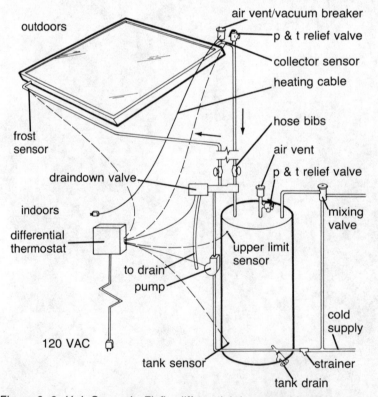

Figure 6–8: H. I. Square's Fixflo differential thermostat has a special frost sensor, in addition to the collector sensor, for detecting quick freezes that might catch other thermostats off guard. Here it's wired to a draindown system. The upper limit tank sensor shuts everything down when the storage tank water gets too hot.

tank. This is an unpressurized, closed loop that's completely separate from the domestic water supply. Water in this loop is pumped from the holding tank up to the collectors and back to the holding tank. A pressurized heat exchanger in the holding tank transfers that heat to the second loop, which is connected directly to the water heater. In one version of this system hot water from the heat exchanger is moved into the water heater only when a hot water tap is opened. A variation on this uses a second pump to pump domestic water between the heat exchanger and the water heater.

Like the draindown system, the drainback also has a differential thermostat and at least two sensors, one at the bottom of the holding tank and the other at the collector outlet. When the collector sensor signals that the collectors are hotter than the water in the holding tank, the pump is switched on. When the water temperature in the holding tank equals the temperature of the collectors, or when the outside sensor detects a lack of solar gain, the pump is switched off, and all the water in the solar loop "drains back" by gravity into the holding tank. As long as the collectors and the solar loop pipes are properly sloped, the collectors will always drain into the holding tank every time the pump switches off. What shall we call it, gravity freeze protection?

Compared with a draindown system, the drainback system may be a little more expensive if you buy a factory-made drainback holding tank and heat exchanger. You can make your own exchanger and put it in an inexpensive metal drum. Also, because the collectors don't directly heat the potable water, some efficiency is lost when heat is transferred through the heat exchanger. The own-

ers of drainback systems usually consider this compromise to be slight in view of the failsafe freeze protection that this system provides.

If you are considering an active system, it would be worth your while to consider the drainback approach (figure 6–9). Its simplicity, reliability and relative low cost (when compared with antifreeze or phase change systems) makes a drainback system a good, safe bet. We like it.

Outside of the hire-it-out approach there are a few paths you can follow to get a drainback system into your house. A do-it-yourselfer can pick up a package of preassembled components in which the pump, differential thermostat, heat exchanger and holding tank come in a single module. The installation of one of these modules can be as easy as connecting four pipes to it, placing the sensors and plugging it in. You can also select and install individual components, which is more time-consuming than installing a module but which can save you a couple of hundred dollars. Last, a more experienced do-it-yourselfer might want to build the holding tank and the heat exchanger, buying only the differential thermostat and the water pump.

Like the collectors in a draindown system, the collectors in this system should be tilted a little so that water will completely drain out of them when the pump stops. The pipe runs leading away from the collectors should also slope continuously downhill, dropping at least ½ inch for every foot of pipe run, and there should be no major ups and downs between the collectors and the holding tank. Because a drainback system has no air vents to help the water drain from the collectors, the tubing in the solar loop should be ¾-inch-i.d. to prevent air locks that might hamper drainage. To save money, ¾-inch-diameter CPVC or polybutylene tubing can be used instead of copper.

The final installation requirement that is unique to drainback systems has to do with the relative positions of the collectors and the holding tank. Because the solar loop is unpressurized, more pumping power is required to move the water around. If there were, say, 40 feet of vertical distance between these two components, a relatively large, and power-hungry, pump would have to be used. In any active system the cost of the energy used for pumping is really a debit against the value of the energy gained from the sun, and the larger the pump the greater this debit becomes. So in a drainback system you want to minimize the vertical separation (feet of head) between the collectors and the holding tank.

This means, for example, that with roof-mounted collectors it's a good idea to put the tank in the attic or in some second-floor closet. With ground-and-wall-mounted collectors the holding tank is often installed in the basement next to the water heater. By keeping the pumping requirement to less than about 20 feet of head, you can get by with a ¹⁄₂₀-horsepower (80-watt) pump and still have an adequate flow rate without negating your solar gains with big electric bills.

Commercial Systems

There are several drainback "packages" on the market that are suitable for do-it-yourself installation. KTA Solar, Inc., of Herndon, Virginia (see Appendix 1), sells the KTA Series 10 DHW System, which incorporates a smart design improvement. KTA recommends that the holding tank be placed no more than 12 feet below the collectors. A ¹⁄₂₀-horsepower pump, plumbed to the drainback tank, forces heated water from the small tank down through connective plumbing leading to a 20-square-foot heat exchanger in the bottom of a stone-lined water heater, which can be installed in the basement. The solar-heated water from the drainback tank gives up its heat in the heat exchanger, then flows back up to the collectors and back down to the holding tank to complete the loop (figure 6–10). This differs from the basic system design in which solar heat is stored in the holding tank, which contains the heat exchanger. Hot water isn't delivered to the domestic supply until a tap is opened. The design variation probably represents a performance improvement, but at the cost of buying a new water heater with an immersed exchanger for at least $400.

The stone-lined 80- or 120-gallon water heater that comes with the KTA system has an auxiliary electric heating element for back-up heating. The heat exchanger in this system is actually coiled inside the back-up tank. (This is called an internal heat exchanger; less efficient solar storage tanks have an external, or "jacket-type," heat exchanger that's wrapped around the outside of the tank.) The heat exchanger itself is a *single-wall* type. (The *double-wall* type is used in antifreeze systems to eliminate the possibility of the antifreeze leaking into the domestic water supply.) Single-wall heat exchangers transfer heat more efficiently from the solar loop to the domestic water than do double-wall exchangers.

In the KTA package the holding tank and the pump make up one preassembled module, and the water heater, differential thermostat, sensors and heat exchangers make up the other.

At the time of this writing, KTA Solar, Inc., quoted a price of "about $1000" for this drainback package with the 80-gallon com-

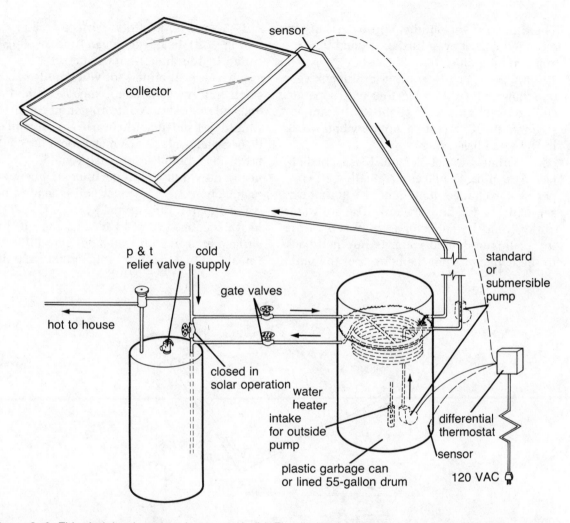

Figure 6–9: This drainback system is easy to build. The drainback tank is made from a 55-gallon drum lined with a heavy-duty polyethylene bag. Ideally the drainback tank is installed as close as possible to the collectors to minimize head pressure. The pump for the solar loop is normally installed outside the drainback tank, but small submersible pumps are available. A submersible pump will add its own waste to the solar water.

bination water heater/heat exchanger. The price is "slightly more" for a package built around a 120-gallon tank.

Adding the cost of 80 square feet of home-built collector area (at a cost of $12.50 a square foot, or $1000; ready-mades come in at $15 to $25) and the cost of connective plumbing, valves and fittings ($200), this drainback system could easily cost $2200 before you're done. You will, of course, be eligible for the 40 percent federal solar tax credit, which would reduce the cost to about $1300.

There is, however, another approach that can reduce this cost even further if you buy separate components.

Figure 6–11 illustrates a drainback system that can be pieced together with separate components. This has a slightly different design that the KTA system, but they both work in the same basic way.

Everhot, Inc., of Boston, Massachusetts (see Appendix 1), makes a 7.5-gallon all-copper heat exchanger tank (model #8) that can serve nicely as a holding tank. The hot water outlet from the collector is plumbed to the top of this tank, and water is pumped back to the collector from the bottom of the tank.

A 17.5-square-foot (total surface area of the tubing coil) heat exchanger (model #6) is coiled inside the all-copper tank.

If you're inclined to, you could make a small holding tank/heat exchanger like this one and save a lot. A 5- to 10-gallon container should be used. It could be a heavy-duty plastic or metal container. A metal container could be waterproofed and rustproofed with a heavy-duty polyethylene liner or fiberglass resin. The heat exchanger coil is made from soft copper tubing (type K). You'll want a heat exchanger with 14 to 20 square feet of surface area, which translates into 60 to 90 lineal feet of ¾-inch tubing, or 90 to 120 lin-

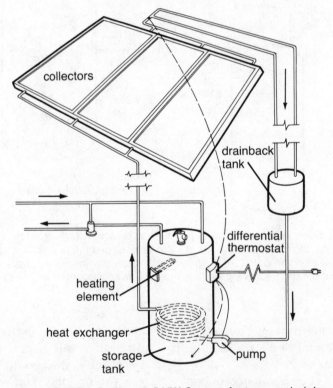

Figure 6–10: The KTA Series 10 DHW System features a drainback tank that is installed in the highest possible heated space to minimize pumping requirements. Water in the solar loop flows down to the basement storage tank, then it is pumped back to the collectors.

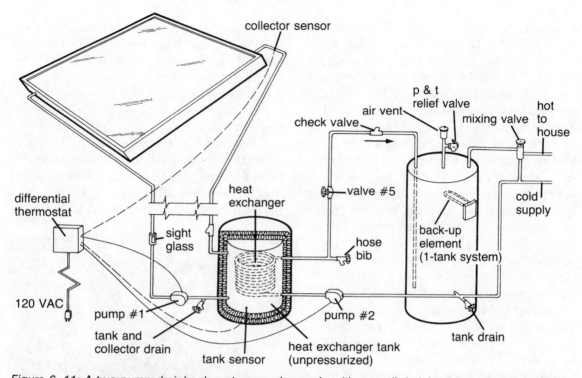

Figure 6–11: A two-pump drainback system can be made with a small drainback/heat exchanger tank. Pump #1, which circulates unpressurized water, is sized according to the drainback pump sizing instructions outlined in figure 6–5. Pump #2, which circulates pressurized water, can be as small as $^{1}/_{100}$ horsepower. The sight glass is used to monitor the water level in the drainback tank.

eal feet of ½-inch tubing. It is probably best to use ½-inch-diameter tubing, as it is easier to work with and has a better surface-area-to-volume ratio, which means better heat transfer. Wrap the tubing around an object with the proper diameter to make a coil that will fit inside the holding tank you've selected. A tube bending tool can make the job a little easier; they're available at plumbing supply stores. This home-built combination is plumbed into the system the same as the Everhot unit. Of course, the tank and pipes should be wrapped with insulation.

In this system water from the conventional water heater is pumped through the heat exchanger by a second pump, which means that solar heat is transferred to the storage tank at the same time that it's being collected. The two pumps in this system are wired in parallel to a single differential thermostat. If the holding tank is next to the water heater, the second pump can be as small as 1/100 horsepower, since there is hardly any head pressure. The horsepower rating of the solar loop pump depends again on the distance between the drainback tank and the collectors and can be determined by reading the information on pump sizing in figure 6–5. Be sure to install the pump below the water line of the holding tank. The outlet from the collector should pour into the holding tank above the water line.

The differential thermostat and the temperature sensors are installed in the standard way, with one sensor at the collector outlet and one at the bottom of the holding tank. The two pumps are wired in parallel to the differential control so that they'll start and stop at the same time. You also have to make sure that the differential control is rated to handle the increased load of two pumps.

The main difference between having the heat exchanger inside the back-up water heater and having it inside the holding tank is one of cost. You can't build your own water heater with an immersed exchanger, and since these units cost at least $400, it may be best for cost savings to go with the holding tank/exchanger combination. The other difference in this system is the use of a second pump that circulates pressurized water between the heat exchanger and the existing water heater. This approach may give an overall performance boost over a single-pump drainback system in which there is no circulation between the holding tank and the water heater. But the numbers aren't in on that one, and there is at least one manufacturer of a single-pump drainback package who believes the second pump is unnecessary.

Chris Fried Solar, of Catawissa, Pennsylvania, manufactures the SunShuttle kit (see Appendix 1), a single-pump draindown system. Because there is no circulation between the holding tank and the water heater, the holding tank must be somewhat larger than the ones in the two previous systems. Since the holding tank in the SunShuttle system essentially stores all the solar heat, it would overheat (and so would the collector) if it wasn't large enough. So Fried specifies a 55-gallon drum lined with a heavy-duty, high-density polyethylene liner. The heat exchanger coil, which is home-built from the directions in his manual, is 60 feet of ¾-inch-i.d. copper. He has found that with a hot water

flow rate (through a faucet or shower head) of 1 to 2½ gpm, the outlet temperature of the coil is within 5°F of the hottest water at the top of the holding tank. This indicates adequate heat transfer to the flowing domestic water, with no need, says Fried, for a second pump.

How do the prices of these latter two systems compare with the $2200 estimate for the KTA-type package? In the two-pump system the Everhot all-copper tank, including the heat exchanger, costs about $300 (at the

Photo 6–8: This drainback tank is a 55-gallon drum with a plastic liner and coiled copper heat exchanger.

time of this writing). Two pumps would cost around $150. A differential thermostat with sensors costs $100 to $150 making a total component cost of about $600. If you made 80 square feet of collectors at $12.50 a square foot, you add $1000, and another $200 is added for pipes, valves and all the other parts. The total comes to $1800, which becomes about $1100 with the tax credit, about $200 cheaper than the single-pump system with the heat exchanger in the water heater.

The SunShuttle kit (single pump, heat exchanger in the holding tank) costs about $380, and it includes some of the basic parts needed to build a 48-square-foot collector (absorber materials, semirigid plastic glazing), the differential thermostat and the plastic liner for the holding tank (55-gallon drum or equivalent). The cost for completing the system with 48 square feet of additional collector area (pump, pipe valves, fittings, more collector materials) could be $670, bringing the total to about $1050, or about $630 when the tax credit is taken.

Three different versions of the same system have a "post-IRS" cost spread of about $600 (from $630 to $1100 to $1300). The performance advantage probably goes to the most expensive system, but whether or not it's a $200 advantage or a $600 advantage isn't really known. The choice thus involves not only performance but initial cost and how well either system integrates with your existing hot water system.

Antifreeze

So far in this chapter we have seen different ways that water is used to capture solar energy. Water has three properties that make it the best all-around solar collection fluid:

It's inexpensive, it's nontoxic, and it captures heat better than just about any other fluid. Water also has one bad property: It freezes at 32°F. Draindown and drainback freeze protection systems are designed to prevent water from freezing in the collector, but some solar engineers have opted to use no water at all, replacing it with an antifreeze fluid, or heat transfer fluid.

The most widely used type of antifreeze are the glycol fluids. Of these, *propylene glycol* is best because it is nontoxic. *Ethylene glycol*, which is automobile antifreeze, is toxic and should not be used. A silicone-based fluid has also been used because it has an extremely low freezing point (under −80°F), but it can be troublesome because it has such a low surface tension that it's been known to leak through sweat-soldered joints. Mineral and even vegetable oils have also been used, but really the verdict is in: Propylene glycol is the preferred fluid for anitfreeze systems.

All antifreeze systems must have two loops. The antifreeze is circulated by a pump through the solar loop, from the collectors to the heat exchanger, then back to the collectors (figure 6–12). The second loop is the domestic water loop, which absorbs heat from the antifreeze through a heat exchanger.

If you were to use a toxic antifreeze solution such as ethylene glycol in the solar loop, common sense (and possibly building code requirements) would require you to use a double-walled heat exchanger to eliminate the chance of an antifreeze leak into the drinkable water. Since it is nontoxic, proplene glycol can be used with a single-walled heat exchanger, which is more efficient (although some local codes require the use of double-wall exchangers no matter what type of fluid is used).

When antifreeze heats up, it expands,

taking up as much as 5 percent more than its original volume. The use of an expansion tank gives the increased volume someplace to go. The expansion tank consists of a bladder of air pressurized to 25 psi, which "gives" or compresses when the antifreeze expands. It also serves to exert a little pressure on the vacuum breaker at the top of the collector loop, which keeps it closed until such time as the loop is drained. The expansion tank is connected to an air purger (figure 6–12), which removes air from the system.

Generally, antifreeze systems are more expensive than their water-filled counterparts, but in return for the added expense you get the security of knowing that it's impossible for your collectors to freeze. This peace of mind doesn't come cheaply. A commercially installed antifreeze system can cost from $3000 to as much as $4500. No matter how you look at it, that's a lot of money to spend on a water heater. By doing your own installation you can, of course, save plenty. There are antifreeze control modules avail-

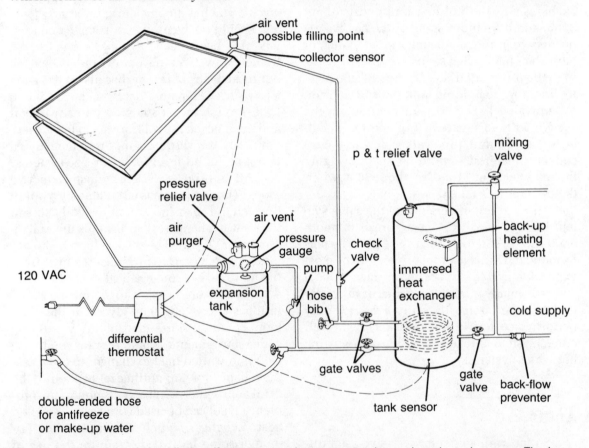

Figure 6–12: A properly installed antifreeze system has an expansion tank and an air purger. The hose at lower left automatically makes up for any collection fluid that may have left the system through the pressure relief valve. If you decide to install such a "make-up" water line it's important to check the mixture of the antifreeze from time to time. Highly diluted antifreeze at some point is no longer antifreeze.

Photo 6–9: This antifreeze system looks simple, and it is. Here you see, left to right, a differential controller, an expansion tank, a pump and a storage tank. The little tank at lower right is filled with the compressed air that's used to fill this system with antifreeze. (A compressed air tank isn't necessary; antifreeze can be poured into the solar loop from the top of the collectors.) The little white pot catches any overflow from the pressure-and-temperature relief valve.

able and, as with the other systems, you can piece together separate components.

Some solar water heating systems should be installed by professionals. There are many simpler systems—like the batch, thermosiphon, draindown and drainback systems—that certainly can be installed by do-it-yourselfers. There are also complicated systems—such as the phase change system described in the next section—that a do-it-yourselfer should never attempt to install. The antifreeze system falls somewhere in between. Less-experienced do-it-yourselfers who want an antifreeze system would be well advised to hire a contractor.

The manufacturers of several packaged antifreeze systems are listed in Appendix 1. You can write to the manufacturers for specifications and prices and also for the name of an authorized dealer in your area. For those who want to go the d-i-y route here are some guidelines: As with any active system, there are no limitations on collector placement vis-à-vis the storage tank. Nor must the inlet be at the low end and the outlet be on the high end, as in a thermosiphon system. Since the collectors don't have to be drained regularly, you don't have to slant them, though it's a good idea because you'll be able to drain them more easily if you have to work on the system or change the antifreeze fluid (every couple of years). When plumbing the system, ¾-inch-i.d. tubing should be used throughout. Either copper or CPVC or polybutylene tubing can be used.

To size the pump, you can again use figure 6–5 to select a pump with the proper horsepower rating for your specific application. In a closed-loop system it's a good idea to install isolation valves on both sides of the pump so that it can be serviced without draining all the fluid out of the solar loop.

The impeller housing of a pump in an antifreeze system can be made of cast iron, since there's no water to cause rusting (although some antifreeze solutions use a mixture of half glycol and half water, which still allows the use of a cast iron pump).

The pressurized water storage tank should be the type with an internal heat exchanger coil (see Appendix 1) that has between 15 and 20 square feet of surface area. One drawback of the glycol fluid is that it can cause corrosion in the plumbing after a few years. To counter that, a nontoxic corrosion inhibitor is added to the antifreeze.

In plumbing the system, some municipalities require a *backflow preventer* where the water main enters the house so that in the unlikely event of a leak the antifreeze won't migrate into your neighbor's drinking water. Check your local plumbing codes.

A float-type air vent should be added at the top of the solar loop, to facilitate both draining and filling of the loop. Make sure that the air vent you get is made for outdoor service, which often means it must be made of stainless steel or plastic. When all the pipes, valves and other fittings are in place the loop is usually filled first with water so leaks can be detected without wasting antifreeze. Then the loop is completely drained and filled with antifreeze through the port meant for the air vent.

You'll never have to test the fluid in a water-filled system, but every six months or so you should drain out a tiny bit of antifreeze and dip some litmus paper into it. The pH should be kept above 6.5. Acidic collection fluid will corrode the solar loop. When the pH drops below 6.5, it's time to drain the solar loop, flush it out with water and then refill it with fresh antifreeze. The antifreeze is usually good for only two or three years and is replaced at cost of around $8 per gallon. If you buy an antifreeze system from a solar contractor (the type most are offering these days) you may also be able to buy a service contract that includes fluid and system maintenance. Also, you may not be using a glycol-based fluid, but a silicone- or hydrocarbon-based one, both of which have different maintenance and check-up procedures. Check the manufacturer's instructions.

Phase Change

Refrigerators, air conditioners and heat pumps all work with a fluid called Freon. In all of these systems liquid Freon absorbs heat and turns to a gas. As the Freon cools in a condenser it gives off heat and condenses back into a liquid. The Freon endlessly "changes its phase," or state, from liquid to gas and back to liquid.

Freon lately has also been used as a solar collection fluid instead of water or antifreeze. Aside from its heat transfer capability, it has another property that makes it attractive for flat plate systems: Its freezing point is around −140°F. Freon also has one bad property: Some say it weakens the earth's ozone layer, which protects us from excessive ultraviolet radiation. This drawback may be enough to dissuade anyone who is concerned about the environment from considering a Freon-filled system. Nothing lasts forever, and sooner or later the Freon will leak into the atmosphere. Since all phase change systems are commercially installed, Freon systems are relatively expensive, costing in the range of $3000 to $4000.

There are presently both passive and active versions of the phase change system. The passive version is similar to a water-filled thermosiphon system. When liquid Freon enters the collector, it absorbs heat and changes to a gas. The hot gas rises and flows to a storage tank above the collectors. The Freon flows through a heat exchanger in the tank, giving up its heat to the domestic water as it condenses, and then returns to the collectors. Active phase change systems commonly have two pumps (figure 6–13). One pump pushes the Freon in the solar loop while the other forces the domestic water through a heat exchanger. The phase change and heat transfer activity is the same as in the passive system.

How well do phase change systems work? In eastern Pennsylvania researchers for *New Shelter* magazine compared an active phase change system with a water-filled drainback system. The phase change system cost more than three times as much as the similarly sized drainback system but only produced half as much energy. Other studies have shown that phase change systems deliver about the same amount of hot water as any other active system. Basically, it's too soon to tell, but in the not-too-distant future there should be enough meaningful data on these systems to allow for some useful comparisons.

Considering the high price, the complicated technology and the environmental controversy surrounding Freon, one would certainly have to pause before rushing out to buy a phase change water heating system. But if you're interested, you'll find some manufacturers listed in Appendix 1. You can write them and ask for the names of some dealers in your area. Before signing any contracts, though, it would be a good idea to get a written guarantee of the thermal performance of the system, which is what you should request for any contractor-installed system.

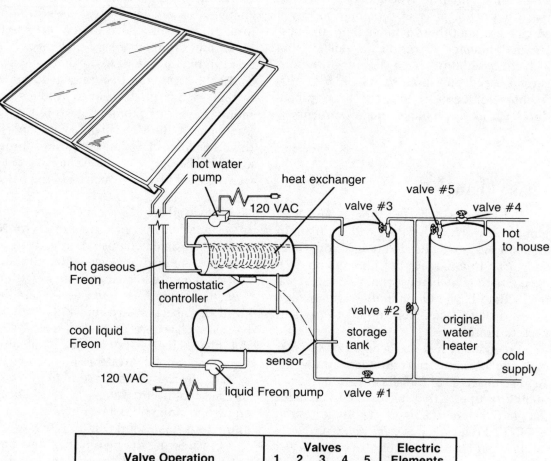

Valve Operation		Valves					Electric Elements
		1	2	3	4	5	
O = open C = closed	solar only	O	C	O	O	C	OFF
	solar & electric	O	C	O	C	O	ON
	electric only	C	O	C	C	O	ON

Figure 6–13: This is a phase change system in its many parts. The valving makes it easy to bring the solar loop into and out of service.

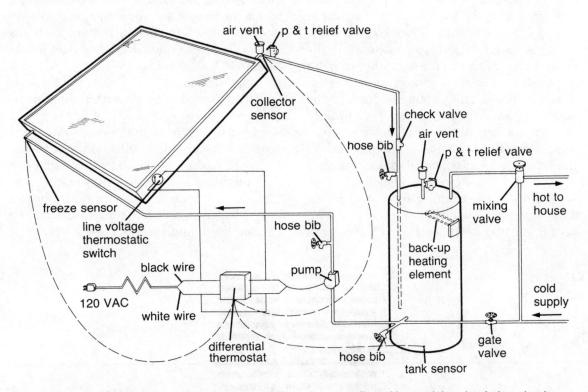

Figure 6–14: A recirculation system requires a differential controller with special recirculation circuitry. The line voltage thermostatic switch shown here backs up the differential controller in case its special circuitry fails to work. It's wired between the power source and the pump so that if the thermostat fails when temperatures drop, the pump will still be switched on to circulate warm water to the collectors. All this works fine unless there's a total power failure.

Other Flat Plate Systems

There are two other flat plate freeze protection systems that are not as widely used as the others but are nonetheless worth mentioning. The application of one system is somewhat limited by climate, while the other system, which uses a nonmetal absorber, is a relatively new development in solar water heating and may turn out to be very widely used.

In warm climates where freezing occurs very infrequently, *recirculation systems* have provided a simple, low-cost means of freeze protection. They are active, open-loop systems that are controlled by the usual arrangement of sensors connected to a differential thermostat that has special recirculation circuitry which is controlled by a third sensor. One sensor is placed at the bottom of the storage tank and the other at the collector outlet. The freeze sensor is placed on the collector inlet. When there is solar energy that can be collected, the differential control switches the pump on and delivers hot water to storage. When the sun is gone and the collector temperature drops to 38° to 40°F, the special recirculation circuitry takes over. In this mode the pump switches on and hot water from the storage tank is pumped through the collectors and the connective plumbing to prevent the formation of ice. This, of course, means that the system is dumping some of the solar energy it collected and stored during the day. If there were constant and prolonged freezing at temperatures well below 32°F, the system would lose too much of its heat protecting itself. This is why recircula-

tion systems should only be installed in the Sun Belt, from Florida to southern California, where the lowest nighttime temperatures are rarely below 20° to 25°F and where the lion's share of winter nights don't even get below 32° to 35° F.

The two main vulnerabilities of this system are a breakdown of the recirculation circuit and a power failure. If there is ever a blackout the same night the temperature approaches freezing the collectors will be unprotected and pipes will burst, as happened to hundreds of systems in California in 1979. There's not much that can be done about a power failure, which further emphasizes the climate limitation on this system. There is,

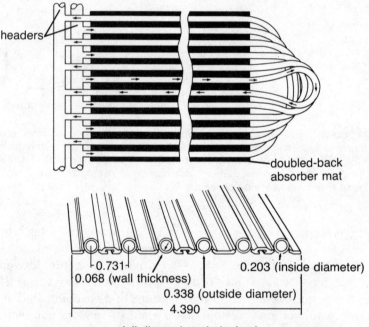

(all dimensions in inches)

Figure 6–15: SolaRoll collectors are made from strips of EPDM tubing mat about 4 inches wide, top. *The mat is stripped away from the tubes, which are inserted alternately into inlet and outlet headers. The mat is doubled over to form an absorber plate,* bottom.

Photo 6–10: SolaRoll is held firmly inside the collector box with mastic.

however, a way to protect the system against a malfunction of the recirculation circuit. As figure 6–14 shows, a line voltage (120 VAC) snap-disc thermostatic switch is installed in parallel with the other switch in the system, the differential thermostat. The snap-disc is set to close when the temperature drops to a present temperature (36°F). When it does close, the pump goes on regardless of what the differential thermostat is doing. Most differential control manufacturers make controllers for recirculation systems. A recirculation differential control is about $10 more than similar controllers without the special circuitry, and you must specify that you want the recirculation function when ordering the controller. The pumps and connective plumbing in this system are sized the same as for draindown systems.

The Rubber Collector

The other type of freeze protection system features a collector absorber material that isn't made of copper but of a synthetic rubber called ethylene propylene diene monomer, or EPDM. Like neoprene, this is an extremely durable material that doesn't oxidize, embrittle or get corroded by anything. It doesn't break down even after 20 years of solar and climate exposure. (This is known because of the long-term service that EPDM has provided as a gasket material, most notably used on the windshields of jet airplanes.)

The developer of the EPDM collector is Bio-Energy Systems, Inc., of Ellenville, New York (see Appendix 1), which calls its product SolaRoll. It is intended to be part of a low-pressure, closed-loop system because the connections between EPDM absorber and the copper headers are only capable of operating at a maximum pressure of 30 psi (figure 6–15). If your water pressure was equal to or less than that, you could actually use a SolaRoll collector in any type of open-loop water-filled system. But since most house water systems run on higher pressure, this collector is best used with a drainback or an antifreeze system. (Bio-Energy makes a drainback module for its collector; see Appendix 1.)

The antifreeze system would of course provide absolute freeze protection, but what if something went wrong with the drainback, and the collector didn't drain? Or what if the collector was part of a low-pressure drain-down system, and the valve got stuck? Well, the collector would freeze, but its rubbery tubes wouldn't burst, they'd just stretch. That's another form of absolute freeze protection.

Sun-Powered Pumps

Today it's possible to power an active solar water heating system without using utility-produced electricity. Photovoltaic cells mounted next to the water-heating collectors convert sunlight to electricity, which powers the pump.

Independent Power Systems of Armonk, New York (see Appendix 1), developed one

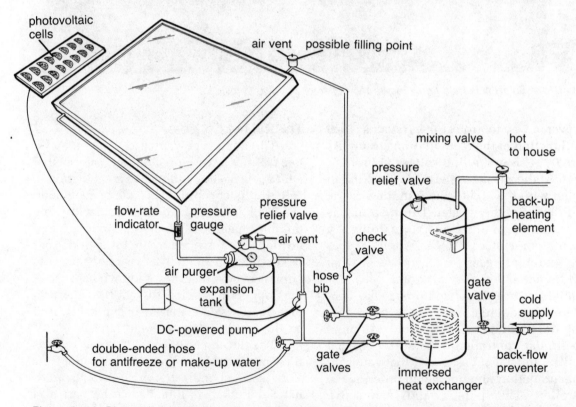

Figure 6–16: Photovoltaic cells convert sunlight to electricity, which powers the pump in this antifreeze system.

Photo 6–11: Photovoltaic cells mounted next to the water-heating collectors eliminate the need for utility power for the pump.

such system. It's basically an antifreeze system without a differential controller. The photo-cell array is wired to a special direct current (DC) pump (figure 6–16). When sunlight strikes the photovoltaic cells, the pump forces antifreeze into and out of the collectors, then through a heat exchanger in the storage tank. When there is no sunshine, the pump stops running. Unfortunately, Independent Power Systems is not presently selling the system, and a company spokesperson said marketing plans have been postponed indefinitely.

The American Solar King Corporation of Waco, Texas (see Appendix 1), currently sells a Photovoltaic Power Pak with its antifreeze system. The price is "about $4000" for the complete, installed system, a company representative said. Even though it's presently possible to buy a photovoltaic-powered system, one gets the impression that all the bugs have yet to be worked out. One problem with the current systems, the Solar King spokesperson said, is that the pump starts at the first sign of sunlight, whether or not there's enough heat in the collectors. In the future you can probably expect to see a differential controller added to these systems so that no fluid is circulated until the collectors have become sufficiently heated.

7

HEATING WATER WITH SOLID FUELS AND HEAT PUMPS

Starting in the mid-1970s, stove sales began to skyrocket as high oil and electricity costs made wood and coal attractive fuels. Coal stoves and woodstoves are not only nice to snuggle around in the winter, but with a little modification they can also heat a lot of water. Usually all it takes is the installation of a heat exchanger in the firebox or in the stovepipe, some connective plumbing leading to the storage tank, a pump, and a differential thermostat. It's also possible to make a thermosiphon-powered stove water heater, which doesn't need a pump. Another recently developed way to heat water is with a heat pump, which takes heat from the ground or air and transfers it to an existing water heater. You can complement your solar water heating system with a stove water heater. A heat pump water heater may be a good way to cut the high cost of electric power, though it may not be more cost-effective than a solar water heater.

Photo 7–1: This stove has a serpentine pipe heat exchanger in the firebox.

Hot Stove, Hot Water

But, wait a minute, you say. You've just invested a bunch of money in a solar water heating system. Why do you need a stove water heater? The answer is clear when you stop to think about the sun. In the summer there's plenty of solar energy available to heat your water, but in the winter there's somewhat less (15 to 30 percent less). Since there's not as much solar energy in the winter, why not add a heat exchanger to the stove (if you already have one) and use it as a winter water heater? In the summer, when you don't need to burn fuel, the sun can take over. In other words, solar and stove water heating are very complementary. A solar/stove combination can actually have a shorter payback period than a solar-only system. If your stove is used for most of your space heating, you could even consider shutting down your solar system in winter; the stove can heat all the hot water you need. Thus a solar thermosiphon or batch system, or any water-filled system for that matter, could be drained to eliminate the potential for a freeze-up.

Stove Heat Exchangers

There are three basic kinds of heat exchangers for stoves: *stovepipe coils*, *serpentine pipes* and *mini-tanks*. The kind you should use depends somewhat on the type of stove you have. Generally, stovepipe coils are used with stoves that are not airtight, such as the old-time potbelly, box or barrel stoves. Such stoves are typically inefficient compared to the more modern designs, and much of the heat of combustion is wasted up the flue. A stovepipe coil is used to capture some of this heat and turn it into hot water instead of Btu's for the cold night air. The other two types are used with the newer, airtight stoves, which are much more efficient at converting

fuel to space heat and don't send as much heat up the stack.

The coil is made with soft copper tubing (½- or ¾-inch i.d., type L) that's been rolled into the shape of a helix to fit inside the stovepipe. The supply and return lines penetrate the stovepipe wall as shown in figure 7–1. The main concern when building or buying a stovepipe coil is not to install one that's too large, or too much heat will be extracted, and the stack will be overcooled. When the exhaust gases are overcooled, they can condense more readily on the upper part of the stack, which can lead to a dangerous buildup of potentially combustible substances (creosote). A hot stove fire could ignite these de-

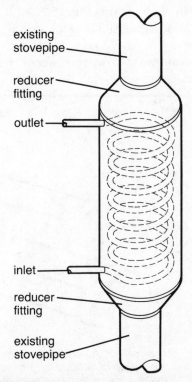

existing stovepipe

reducer fitting

outlet

inlet

reducer fitting

existing stovepipe

Figure 7–1: A stovepipe heat exchanger coil assembly is added to the existing stovepipe with reducer fittings.

posits and cause a chimney fire. It's important, then, that you not exceed the design guidelines for a stovepipe coil. It's already sized to deliver a lot of hot water.

These days stoves are made to give the owner much more control over the amount of air that enters the firebox. The less air, the slower and more efficiently the fuel is burned. Since there is less heat and a slower airflow going up the stack of an airtight stove, there is more potential for creosote deposition inside the stovepipe. Extracting stack heat for hot water would only exacerbate the problem, and this is why either a serpentine pipe or a mini-tank is used inside the firebox of an airtight stove.

The serpentine heat exchanger consists of several lengths of straight steel pipe joined by 180-degree elbow bends (figure 7–2). It is usually installed along the back or side wall of the firebox. You can make a serpentine unit that's custom-sized to your stove with parts that are commonly available.

A mini-tank heat exchanger, which is also installed along the side or back wall, is a flat, rectangular container with inlet and outlet pipes on the outside and a baffle on the inside (figure 7–3). Since they require metal cutting and welding, these are probably the most difficult heat exchangers for do-it-yourselfers to make, but you can buy one through the mail (see Appendix 1).

To install a mini-tank or serpentine unit, you have to be willing to drill two holes through the side or back of your stove for the inlet and outlet pipes. The maximum size is 1 inch in diameter, and the holes are usually made at the rear end, so you won't be marring your favorite heater to any great extent.

Stove Systems

Like solar water heaters, a stove system can either be active or passive. A passive stove water heater shares many of the same advantages and disadvantages of a passive solar water heater. The advantages: no pumps, no electricity, and therefore fewer things that can go wrong. The disadvantages: the storage tank must be located above the stove and the connective plumbing should be short, straight

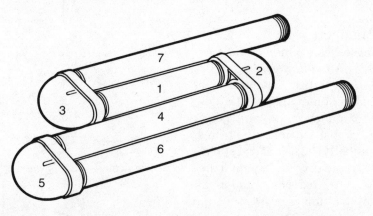

Figure 7–2: A serpentine heat exchanger is made with sections of steel pipe of various lengths and elbow joints. The numbers indicate the order in which the parts are assembled. A serpentine like this is usually installed inside the firebox of an airtight stove.

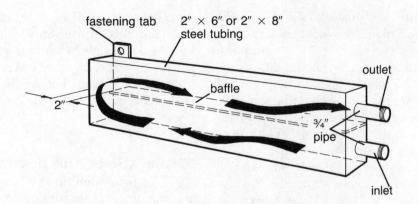

Figure 7–3: A mini-tank heat exchanger can be made from 2 by 6- or 2 by 8-inch steel tubing. The baffle encourages maximum heat absorption. Like the serpentine, it is usually installed in the firebox of airtight stoves; a fastening tab welded to its side makes installation easier.

and simple, with no ups and downs in the pipe runs. The tank should be near the stove, no more than 50 feet away, and the connective plumbing should be 1-inch-diameter copper or steel pipe to encourage easy flow.

Active stove systems, like their solar counterparts, are more adaptable. The storage tank can be placed virtually anywhere, and the connective plumbing can snake every which way between the stove and the tank. Either ½- or ¾-inch-diameter pipe (copper or galvanized steel) can be used.

Making a Stovepipe Coil

You'll need 15 to 20 lineal feet of copper tubing to make a heat exchanger that's adequate for most hot water needs. A longer coil will extract more heat, but if the heat exchanger is too long, too much heat will be removed from the exhaust gases and promote creosote deposition. Thermosiphon systems require a coil made from ¾-inch-i.d. tubing. The coil itself shouldn't impede the flow of exhaust gases, so the inside diameter of the coil should be as wide as the existing stove-

pipe. For example, if you have a 6-inch-diameter stovepipe, the tubing should be wrapped around a 6-inch-diameter form (such as a section of the stovepipe). The coil is then encased in a short section of 8-inch-diameter stovepipe, which is joined to the existing stovepipe with standard reducer fittings.

Fill the tubing with sand to help prevent kinking while you're wrapping the coil, and tape the ends of the tube so the sand won't leak out. Find a proper-sized cylindrical object to form the coil. Stovepipe will work (it's certainly the right size) but it's been known to squash and distort the coil's helical shape. To get around that problem people have used large-diameter PVC or steel pipe, and even rounded logs. The easiest way to make the coil is to place the form in a vise and get a friend to help. Start by straightening the copper tubing you'll be coiling. Needless to say, the heat exchanger coil will get pretty hot when it's placed inside the stovepipe. For this reason you should leave about 16 inches of straight tubing before you start wrapping the coil around the form. The heat exchanger will eventually be soldered to the connective plumbing, and the solder joint could melt if it was much closer than 16 inches from the stovepipe. (In any event, you should use silver solder, which has a higher melting point than regular solder.) Wrap the tubing slowly around the form. Leave another 16-inch straight section at the other end. When you're through bending, pour out the sand.

Take out a short section of your existing stovepipe 12 to 24 inches above the flue collar. To insert the coil in the larger stovepipe, open the stovepipe at its seam. Punch or drill two holes in the stovepipe (⅝-inch holes for ½-inch tubing, ⅞-inch holes for ¾-inch tub-

ing), insert the coil ends through the holes and then close up the stovepipe around the coil. Seal any gaps between the holes and the copper tubing with stove or furnace cement. This assembly can then be fitted to the existing stovepipe with the reducer fittings (figure 7–1).

Making a Serpentine Heat Exchanger

This is another do-it-yourself exchanger you can make with a little help from a plumbing or hardware store to cut and thread the steel pipe sections. You'll notice in figure 7–2 that you need different lengths of pipe to put this exchanger together with 180-degree return bends. Size the serpentine according to the proportions of the wall inside the firebox. The side wall is most often the best place. In box-type stoves the back wall is often the coolest part of the firebox (figure 7–4). Four lengths of pipe (connected with three return bends) are usually adequate for hot water production. Large stoves can accommodate serpentines made from as many as six courses of pipe, but don't get carried away. The stove won't be as effective for space heating if one of its interior walls is completely covered by the exchanger.

After you've decided where you'll install the heat exchanger, lay out the longest length

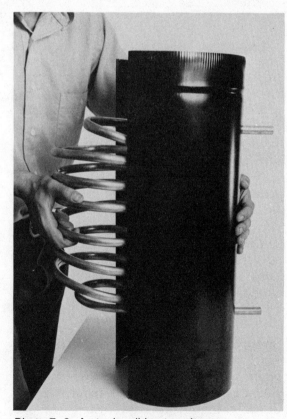

Photo 7–2: A stack coil heat exchanger.

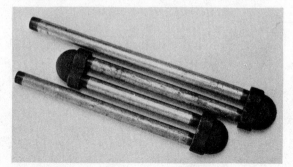

Photo 7–3: A serpentine heat exchanger.

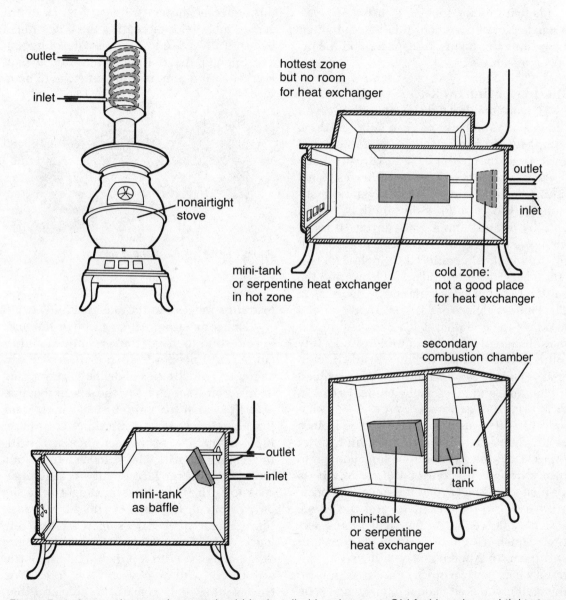

Figure 7–4: A stove heat exchanger should be installed in a hot spot. Old-fashioned, nonairtight stoves send lots of heat up the stack; a stovepipe coil is a good choice for these stoves. Serpentine and mini-tank heat exchangers are placed in the firebox hot spots of airtight stoves. Mini-tanks can be used as heat-capturing baffles that slow the flow of hot air from the stove. Such baffles heat plenty of water, and they can also improve the space-heating capability of the stove.

of pipe first, and then the shorter lengths. Join the threaded lengths of pipe to the return bends with pipe dope and use two pipe wrenches, or one wrench and a vise to tighten everything up. Figure 7–2 shows you the order of assembly.

Making a Mini-Tank

If you're able to do some welding you can make a mini-tank. You can build the thing completely from scratch with plate steel, but you can get four of the six sides already made if you buy a piece of what is known as "steel tubing," which is available in several dimensions. A good size for a mini-tank is 2 by 6- or 2 by 8-inch tubing that can be 10 to 24 inches long. The minimum wall thickness of the tubing itself should be 1/8 inch. The two end pieces, the inlet and outlet pipes and the baffle are arranged as shown in figure 7–3. The baffle is a piece of flat stock that's welded to one of the end pieces. It's 2 inches shorter than the length of the steel tubing, and a tiny bit narrower than the tubing's inside dimension. The inlet and outlet pipes are welded to the same end piece as the baffle. Use standard black, not galvanized, pipe. A $^{15}/_{16}$-inch hole is drilled at each end of the end piece to fit the ¾-inch-diameter inlet and outlet pipes. These pipes should be long enough to pass through the wall of the stove with a 16-inch extension beyond the stove. The pipes are welded to the end plate, and the whole assembly is welded to the steel tubing section. Finish the job by welding on the other end plate. In Appendix 1 you'll find an address where you can order a mini-tank through the mail if you aren't going to make one. You can also job the work out to a local welding shop.

The mini-tank is usually installed along one of the walls of the firebox, but if your stove is unbaffled, and heat passes unabated up the stack, you can consider installing the mini-tank as shown in figure 7–4, so that it acts as a baffle, intercepting the flow of heat before it disappears into the wild blue yonder. Such an installation can produce lots of hot water and a more efficient stove to boot.

Photo 7–4: A mini-tank heat exchanger.

Installing the Serpentine and the Mini-Tank

Holes must be drilled through the wall of your stove to install the serpentine or mini-tank. You'll need a ½-inch drill, which can be rented, and a series of drill bits graduating in size from 1/8 inch up to the size of the final hole ($^{15}/_{16}$ inch for ¾-inch steel pipe). Mark the holes by lining up the inlet and outlet pipes of the heat exchanger, then start drilling. Hold the drill with a loose hand when you start using the larger drill bits as it might start to spin if the bit gets caught on a big burr of metal. (Spinning drills hurt hands.) Sheet metal stoves are somewhat easier to work with. You can drill a small hole and then enlarge it with a punch. Plate steel can also be cut with an oxyacetylene torch. You can use the torch to start the hole and then finish it off neatly with a drill bit.

The serpentine or mini-tank should be supported inside the firebox. A small tab or a metal strap can be welded to the exchanger, and this in turn can be attached to the wall of the stove with a small stove bolt. After everything has been securely installed, seal up any air gaps with furnace or stove cement.

Tank Sizing

Unlike solar water heating systems, the thermal output of stove water heaters is dif-ficult to predict. Solar gain is predictable from years of accumulated weather data, but stove use patterns vary widely from household to household. You'll probably want to buy a large storage tank if you're going to heat water with both the sun and a stove, but if you're going to heat water with just a stove, you can use

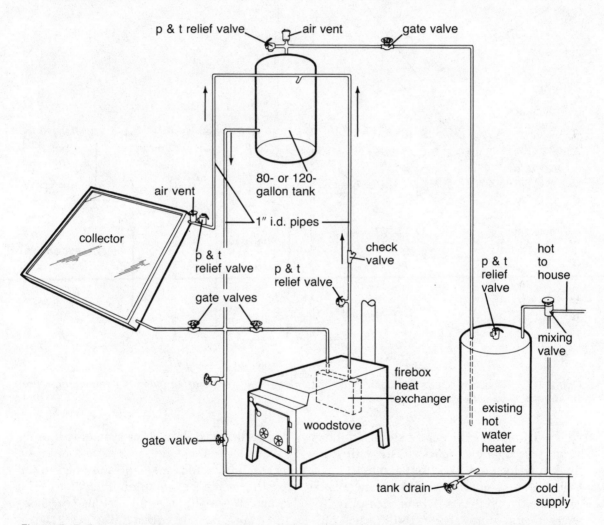

Figure 7–5: A passive solar/stove combination heats water with renewable fuels all year. The solar loop is valved off before the first frost, then returned to service in the spring, at the close of the wood- or coal-heating season. Mixing valves, which are standard equipment in all solar water heating systems, are particularly important in stove systems.

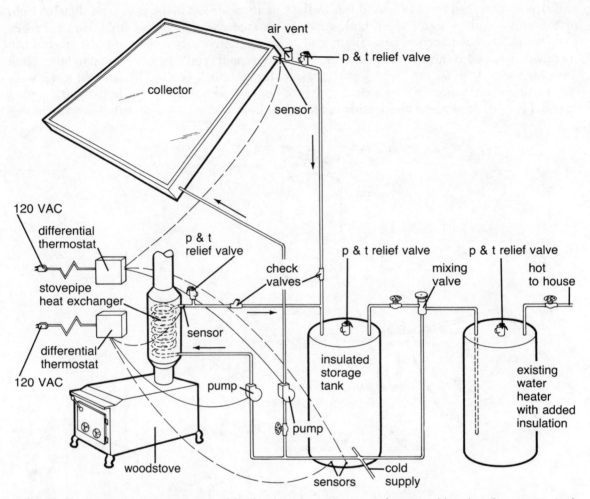

Figure 7–6: This active solar/stove combination has the advantage of not requiring that the storage tank be above the collectors or the stove.

your existing water heater for storage if it has at least a 52-gallon capacity. This will store the hot water produced from about 10 to 14 hours of stove use a day. If the storage tank consistently overheats (which means an average temperature above 150°F), the efficiency of the system will be increased if you get a larger storage tank, such as an 80- to 100-gallon unit.

A thermosiphon stove system will probably require a second storage tank anyway, located above the stove and plumbed to the existing water heater in another part of the house. Buy a high-quality, stone- or glass-lined tank, rated to withstand at least 125 psi.

A combination solar/stove water heating system will require a larger storage tank. If it's equipped with a back-up heating ele-

Photo 7–5: This stove has a mini-tank exchanger in the firebox.

ment, this tank can eliminate the need for the existing water heater. Forty to 60 square feet of collector, coupled with 14 hours of wood-burning a day, calls for at least an 80-gallon storage tank. Larger collector area and/or more hours of stove use dictates a 100- to 120-gallon tank. If you've bought the largest tank you can find and it's still overheating, place it in a spot where its heat loss will be a space heating gain.

Figures 7–5 and 7–6 illustrate a solar/stove thermosiphon and a solar/stove active system. If you want to install just the stove water heater, ignore the solar loop, and everything else stays the same. Don't use plastic pipe in either stove system; it will droop with the high temperatures.

You can probably get by with a $\frac{1}{100}$ to $\frac{1}{50}$ horsepower pump in the active system, since much head pressure won't be needed to move water from a basement storage tank to a stove on the first floor (see the pump sizing instructions in figure 6–5). All active stove systems should have a pressure-and-temperature relief valve near the heat exchanger outlet. If the pump or the differential thermostat fails, the valve will relieve dangerous pressures. It's a good idea not to put any shutoff valves between the heat exchanger and the relief valve, as their seals might be damaged by high pressure or temperature. The drain line from the valve can be plumbed to some out-of-sight place, such as the basement or a crawl space. Install the relief valve at least 18 to 24 inches away from the stove to prevent radiant stove heat from setting it off.

You should also install a mixing valve on the hot water outlet of the storage tank regardless of whether the stove heater is active or passive. Stove water heaters have been known to raise the temperature of water to 200°F. The tempering valve will make sure no one is scalded. Use high-temperature silver solder for all the connections near the stove. Otherwise the solder might melt when temperatures get high.

The Advantage of a Solar/Stove Combination

As was mentioned, it's hard to predict how much hot water you'll get from a stove. Anywhere from 40,000 to 150,000 Btu can be put into a hot water tank with eight to ten hours a day of stove use. Much of that output, of course, depends on such things as the intensity of the fire and the design of the stove and the heat exchanger.

A stove water heater working in concert with a solar heater can actually shorten the payback period to less than the payback on a solar-only system. For example, suppose you had 40 square feet of solar collector that produced, on the average, 500 Btu per square foot every day, or 7,300,000 Btu a year. If the solar heater is replacing electricity that costs 7¢ per kilowatt-hour, this means that about 2139 kilowatt-hours of electricity will be saved annually worth about $150. If the solar water heater costs $1000, and a 40 percent federal tax credit is claimed, the true price of the system is $600, and it has a payback period of a little over 4 years (this is assuming electricity prices won't increase faster than inflation in those six years; higher electricity costs will mean an even shorter payback period). Let's further suppose that you add a $350 stove heater, which shares the storage tank with the solar heater. You'll want to use your stove for about 150 days in the winter. Let's say it delivers 40,000 Btu a day for water heating, or 6,000,000 Btu over the course of the woodburning season. That equals 1758

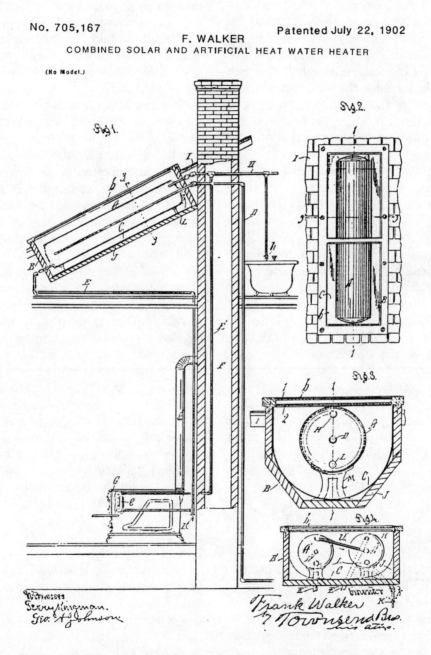

No. 705,167   Patented July 22, 1902

F. WALKER

COMBINED SOLAR AND ARTIFICIAL HEAT WATER HEATER

(No Model.)

Fig.1.   Fig.2.

Fig.3.

Fig.4.

Figure 7–7: This rendering of a passive batch/stove combination is from a turn-of-the-century patent application. Good ideas aren't always new.

kilowatt-hours of electricity, or $123 worth of energy at 7¢ a kilowatt-hour. This combined solar/stove system costs $950, and it saves $273 worth of electricity annually, for a payback period of 3½ years. The woodstove heater costs a little more than half as much as the solar heater (after the tax credit is taken), yet the output of the combined system is 80 percent greater than the solar heater alone. The stove, in effect, lowers the cost of solar energy.

Many people report that it pays to heat water in a stove. Richard Conrat, of Healdsburg, California, is a solar contractor who owns a solar/stove system and has done a great deal of government-funded research and development on these systems. "I've found that woodstove water heating is economical if you burn at least two cords of wood a year," he reported. Conrat has a draindown solar system with 90 square feet of collector. The solar heater shares a 120-gallon storage tank with the woodstove. Six months out of the year, in the summer, the draindown system supplies about 75 percent of Conrat's hot water. For three months in the dead of winter the stove alone heats the water. The only time the back-up electric heater is used, Conrat said, "is on those oddball spring and fall days when it's too warm to burn wood in the stove and too cloudy for solar heating."

Heat Pumps

Webster's dictionary defines absolute zero as "a hypothetical temperature characterized by complete absence of heat and equivalent to approximately $-273.16°C$ or $-459.69°F$." Everything above absolute zero has some heat in it: the coldest air or groundwater, even ice contains a little bit of heat. A heat pump is a device that extracts that heat and uses it to heat water or air.

It is very common these days to hear heat pumps described as "technological wonders," or the "solution to tomorrow's energy needs." But before we accept the claims, let's get to the bottom line. A 1981 report by the Solar Energy Research Institute (SERI—a U.S. Department of Energy contractor) concluded that heat pumps for domestic water heating would be more cost-effective than solar energy if the solar water heating system cost more than $3800 and delivered less than 45 percent of the energy needed for water heating (at a *real discount* rate or interest rate after inflation of 3 percent, which is a common residential rate these days). If the interest rate after inflation was 10 percent (as it is for many commercial situations), the solar water heater must cost more than $2700 and provide less than 60 percent of the energy needed for a heat pump to be more cost-effective. This assumes that the money to pay for either system is being borrowed.

In simpler words, so long as a solar water heating system costs less than about $2700 and delivers at least a 60 percent solar fraction—and nearly all the solar heating systems described in this book, except perhaps the phase change system, meet these requirements—you'll be better off using conservation and a solar water heater.

How They Work

There are six basic kinds of heat pumps. One type extracts heat from outside air and transfers it to the air inside the house for space heating. This is an *air-to-air* heat pump. Another type does the same thing but also heats domestic water. This is an *air-to-air-and-water* heat pump. Another design extracts heat from indoor air and uses it to heat domestic water. This is an *air-to-water* heat pump. Still another version takes the heat from groundwater and uses it for space heating. This is—you guessed it—a *water-to-air* heat pump. Another groundwater design heats domestic water as well as the living space. This is a *water-to-air-and-water* heat pump. The sixth type is a *water-to-water* heat pump, which draws heat from groundwater to heat domestic water.

Each type uses Freon as the heat transfer medium. The water-to-water system, for example, pulls water from a well and pumps it into a storage tank. Liquid Freon is forced through a heat exchanger in the tank, where it absorbs heat from the well water and vaporizes. The gaseous Freon is compressed, then is channeled through a second heat exchanger that is immersed in a domestic water tank. The gas releases its heat to the water, then condenses back into liquid Freon, and the cycle is repeated. Units that extract heat from air work the same way, except that the liquid Freon is vaporized by air, rather than water.

Heat pumps relying on outside air as the heat source are most efficient in warm, southern climates. In places where the outside air temperature drops below 45°F the heat pump must work very hard to extract heat. This means that air source heat pumps shouldn't be used in the colder regions of the United States. Water-source heat pumps, on the other hand, require groundwater that's at least 50°F. It turns out that the groundwater below most of the continental United States is this warm or warmer, and therefore suitable for water-source heat pumps. Does this mean that water-source heat pumps are an answer to the need for lower-cost water heating? It depends on how you look at it.

Are They Worth It?

Promoters of water-source heat pumps are fond of referring to groundwater as "free fuel." But the heat in groundwater isn't really free for the taking any more than solar heat is free. The advantage of heat pump water heating is that the heat pump is more efficient than standard electric resistance heating.

A conventional electric water heater is said to be 95 to 99 percent efficient in converting 1 Btu of electrical energy into 2 Btu of heat. Heat pumps, on the other hand, might produce 2 to 3 Btu's of heat for every Btu of electricity used to run the system. This means that if you were to replace your electric water heater with a heat pump, your water heating costs might be reduced by one-half to two-thirds. Since electricity is the most expensive home energy, that's certainly a significant reduction. But again, that's compared to electric, not solar, water heating.

To fully evaluate the difference, you can look at how the savings of heat pump water heating translates into a payback period on the cost of a system. The least expensive heat pump water heater is the kind that extracts heat from indoor air. This type costs between $700 and $1000 and can be installed by a handy do-it-yourselfer. The heat pump unit is added next to the existing water heater and is connected to it by two plumbing connections (figure 7–8). The heat pump does take heat from this space, but since the water heater is probably located in the basement or in a cooler utility room, it won't increase your space heating load by the amount of heat it extracts.

If a family of four uses 15 kilowatt-hours (kwh) per day (5475 kwh per year, or 18.7 MBtu) for water heating, at a cost of 7¢ per kwh, the annual water heating bill would be $383. A heat pump with a coefficient of performance of 2.5 (2.5 Btu delivered for 1 Btu of electricity used) would use 6 kwh per day to provide the same amount of hot water. This would cost $153 per year, $230 less than the cost of running the resistance water heater. Thus, the cost of the heat pump would be paid back in about four years, and the cost of a kilowatt-hour goes from 7¢ to 2.8¢. It can be said, then, that "heat-pumped" hot water competes favorably with an electric water heater, but as the SERI study concluded, it doesn't win out against conservation and solar water heating.

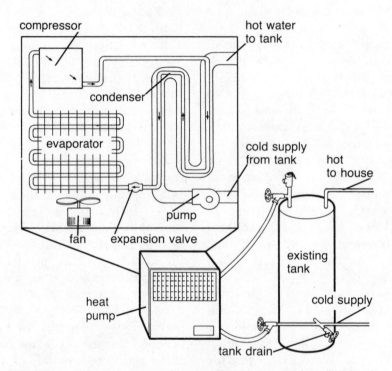

Figure 7–8: Heat pump water heaters like this can be installed by do-it-yourselfers. This unit draws heat from room air.

Photo 7–6: A DHW heat pump connected to an existing water heater.

8

HEATING SWIMMING POOLS, HOT TUBS AND SPAS

A bad tooth had brought me to my dentist, and while waiting for the Novocain to take effect we made small talk. What was I doing these days? I was writing this book. No kidding, he said, and he went on to explain that he was installing a swimming pool at home. Did I know anything about solar pool heating? My mouth was going numb. Yes, I replied. Marvelous, he said. Perhaps I could fill him in. (He was standing there holding the drill. I was stalling for time.) He said he had just bought a $400 propane pool heater, and he figured that heating his pool with gas was going to be expensive. My tongue was getting heavy from the Novocain, but I told him about conservation. It doesn't make any sense to heat a pool with gas, oil or solar energy only to have it escape into the air at night, I explained. That's why your best first step is to buy a swimming pool cover and employ other conservation measures.

The Energy-Efficient Swimming Pool

You might be surprised to know that fully 95 percent of the heat loss from a swimming pool occurs right from the surface of the water. Evaporative heat losses alone account for 65 percent. As liquid water is evaporated, it takes a lot of heat with it. Convective heat losses account for another 20 percent of the total, and radiative losses add another 10 percent.

It's plain to see: Swimming pools need covers. The benefits of using a cover are about as remarkable as the amount of heat lost at the surface. To find out what a cover would save, energy expert Harry Sigworth, Jr. did some extensive analyses using a computer program he helped to develop at the Lawrence Berkeley Laboratory in California. His results would seem to make pool covers worth their weight in gold.

In his work he modeled the energy use of a standard-size, outdoor, in-ground pool in seven locations covering all climate areas of the United States. Table 8–1 shows, for example, that if somebody in Washington, D.C., wanted to use their pool from April to October (keeping the water at a toasty 80°F), they would be facing a heating load of 4.1 therms of natural gas (100,000 Btu/therm) for every square foot of pool surface area. But look what a pool cover does: It reduces that load by 71 percent!

Let's play with some money numbers: Assuming that the pool measures 20 by 40 feet and 1 therm costs 75¢, the cost of heating the pool from April to October would be: (therms/ft²) × (pool area) × ($/therm), or 4.1 × 800 × $.75 = $2460. Enter the pool cover. It cuts the heating load by 71 percent, down to 1.2 therms per square foot for the same heating period. Plugging in that load, the equation becomes: 1.2 × 800 × $.75 = $720. In one year $1740 is saved, and the cover

TABLE 8–1

Energy Savings with Cover and Collectors

City	Pool Heating Season			
	June to August	May to September	April to October	All Year
Yuma, Ariz.				
Standard pool heating requirements*	0	0.3	0.8	7.4
Savings with cover	0%	100%	100%	64%
Savings with cover and collectors	0%	100%	100%	89%
Atlanta, Ga.				
Standard pool heating requirements	0.1	1.2	3.1	11.8
Savings with cover	100%	100%	77%	58%
Savings with cover and collectors	100%	100%	97%	65%
Washington, D.C.				
Standard pool heating requirements	0.4	1.7	4.1	14.1
Savings with cover	100%	82%	71%	54%
Savings with cover and collectors	100%	100%	100%	60%
Los Angeles, Calif.				
Standard pool heating requirements	0.8	1.8	3.2	9.1
Savings with cover	100%	89%	84%	66%
Savings with cover and collectors	100%	100%	88%	87%
Sacramento, Calif.				
Standard pool heating requirements	0.9	2.2	4.0	11.2
Savings with cover	100%	86%	78%	56%
Savings with cover and collectors	100%	100%	98%	69%
Denver, Colo.				
Standard pool heating requirements	1.8	4.1	7.2	18.2
Savings with cover	100%	81%	71%	58%
Savings with cover and collectors	100%	95%	88%	65%
Milwaukee, Wis.				
Standard pool heating requirements	1.7	4.0	7.2	20.3
Savings with cover	88%	75%	61%	49%
Savings with cover and collectors	100%	93%	75%	56%

*The "heating requirements" are the amounts of energy needed to maintain a pool at a comfortable swimming temperature—the number of therms (100,000 Btu) needed per square foot of a pool without an insulating cover. To convert therms to actual fuels, multiply by 1.10 for gallons of oil; by 1.33 for cubic feet of gas; by 1.46 for gallons of LPG (propane); and by 29.30 for kilowatt-hours of electricity.

Source: The information in this table was derived from computer studies carried out by Harry W. Sigworth, Jr. based on "Pools" Computer Program developed at Lawrence Berkeley Laboratory in California. Reprinted from "Solarized Swimming" by Frederic S. Langa (*New Shelter*, April, 1982).

Note: The "savings with cover" figures assume that the pool is protected by an insulating foam cover 0.1 inch thick from 6 P.M. to 8 A.M. daily; the water is maintained at 80°F day and night; and the pool is in a sunny but wind-sheltered location. The "savings with cover and collectors" figures additionally assume that the collector surface area is equal to pool surface area. Larger or smaller collectors will yield different savings.

itself might cost between $240 and $1000, depending on what kinds of accessories are bought with it (manual or automatic roll-up and storage devices).

If you're now champing at the bit, car keys in hand, ready to race over to the pool cover store, please pause for a while. There are several kinds of covers now on the market, and you ought to know what's what before taking the plunge.

There are three kinds of sheet-type covers and couple of unusual variations. They are all made of materials that resist the decay that could be caused by solar radiation (specifically ultraviolet radiation) on a plastic.

That's a standard feature. One sheet type is simply a single layer of a heavy-duty plastic. This kind will primarily cut evaporative losses and have some insulating value, but the two other types will insulate better. One is a sheet of foamed plastic that is around 1/10 inch thick. The material is tough and very tear resistant. The third type consists of two layers of plastic that are joined so as to form larger air bubbles. The result looks like the bubbly packing material that serves as a shock absorber for things that are shipped in boxes. Of these two, the foam type has the advantage of taking up less storage space when it's rolled up.

Photo 8–1: This pool cover is pulled across the pool by hand.

Photos 8–2, 8–3: Pool covers don't have to be in the way when they're not on the pool. In the system shown here the cover returns to its in-ground storage after being deployed by a lightweight, rolling boom that spans the pool.

Speaking of rolls, there are all kinds of roll-out/roll-up systems available to minimize the hassle of dealing with hundreds of square feet of cover material. With small pools of 300 square feet or less you could get by with a totally manual system. Two people could handle the on/off routine. Bigger covers, however, really should be handled with a roller. Some rollers are fastened at one end of the pool, while others roll over the pool as the cover is being let out or taken in. The high-ticket system includes a motorized roller and side tracks that keep the cover stretched across the pool. The tracks are also claimed to be a safety feature: If someone fell into the pool with the cover on, it would just hold them up. (With *any* type of cover *never* take a dip unless it's been completely removed, and the cover itself should either be totally on or totally off at any time.)

An aesthetically interesting but less efficient variation on the sheet cover is the lily pad, not the kind frogs use, but large, 3-foot-diameter circles of plastic that float on the surface. They're easier to manage because they are handled one at a time, but because of their shape they don't give total coverage of the pool surface. Another variation is unique, to say the least: A few thousand plastic foam balls the size of oranges can be spread over the pool and just left there. They're so light that they don't get in the way or konk you on the noggin when you dive in. Like the oversized lily pads, though, they don't provide 100 percent coverage.

Other ways to save include lowering the thermostat on the existing pool heater. Keeping the pool at 80°F is keeping a pretty warm body of water, but for every 1°F reduction, pool heating energy use is reduced by 5 to 10 percent. You should also turn off the heater completely for nonuse periods as short as one night (if you have a pool cover). Some pool owners use a timer to automate the on/off control of the heater. Another heat loss reducer involves protecting the pool from wind. Windbreaks in the form of fencing, bushes and trees give privacy and keep breezes from chilling the poolside crowd. They also reduce evaporative heat losses when the cover is off. If you're about to put in a new pool, you should plan for windbreaks and also go for a sunny location over one that is more shaded. Last, you might be attracted to the idea of painting your pool black or some dark color to increase the absorption of solar en-

ergy. It will cause an increase, but not much of one. Studies show that less than 7 percent of the available solar energy actually reaches the floor and lower sides of a pool, and painting it black would result in less than a 10 percent reduction in pool heating costs. But since the water itself does absorb most of the energy, you could improve collection by using the transparent bubble-type cover instead of the opaque foam plastic.

The Solar Swimming Pool

Before you go further into a solar pool heater it might be a good idea to see what kind of heating bills you get after employing the various conservation measures. If you're not aiming to extend your swimming season into the cooler months (April, October), conservation alone could give you enough of a savings. You can be certain though, that if you want to minimize your gas, oil or electric pool heating with a solar system, the solar investment will be worthwhile.

As table 8–1 shows, a properly sized solar pool heater will provide, in all but the coldest location (Milwaukee), at least 88 percent of a pool's heating needs for the April-to-October period. (Remember that these percentages are for pools kept at 80°F. A reduction of 2° to 4°F would probably mean 100 percent solar heating in all locations. You might only have to make that reduction in the two "fringe" months, April and October, to get that 100 percent.)

Pool heating is probably one of the best applications for an active solar system in terms of both the energy output and the payback on the investment. Only a small temperature rise (5° to 10°F) is needed to make a cool pool a comfortable one, and pool heating collectors are perfect for producing that kind of

Photo 8–4: Pool insulation comes in many shapes and sizes, although the gaps between these "lily pads" make this type somewhat less effective than a full-coverage type.

increase. In fact, pool collectors don't even have to be glazed to do that; they're really just unglazed, unboxed absorber plates placed in the sun.

The absorbers themselves are usually made of plastic or synthetic rubber. A copper plate could be used, but could suffer from the corrosive effects of chlorinated pool water. The plastic and rubber types are cheaper and do a completely adequate job, though they, too, can be degraded by the ultraviolet radiation in sunlight. An aluminum absorber should never be used, as it will corrode quickly when exposed to pool water. Most pool builders, as well as solar contractors, offer solar pool heating systems, so you shouldn't have to go too far to find a few different brands.

The Fafco Company, of Menlo Park, California (see Appendix 1), is a leading man-ufacturer of pool collectors. In mid-1982 a 4 by 10-foot absorber cost about $200, and a 4 by 8-foot unit about $175. The panels can also be bought with glazing, which can extend the swimming season a few weeks in the spring and fall. But the most cost-effective approach is with the unglazed absorber.

Another kind of collector material is the synthetic rubber EPDM, which carries the SolaRoll brand name (see chapter 6 for more information on this product). This flexible, absorber-on-a-roll material can be connected to headers in various configurations, and costs about $5 per square foot. (Typically, pool collectors cost from $3 to $10 per square foot, and the additional hardware and controls can add another $1 to $3 per square foot.)

The workings of a solar pool system are quite simple. The solar loop can, in most situations, be easily spliced into the existing pool heating/filtration system. When the pool pump is on in the existing system, the water is drawn from the pool and passes first through a strainer. Next comes the pump, and then the filter and the conventional heater. From the heater the water returns to the pool. A thermostat in the heater decides whether or not the water is warm enough. If it isn't the heater is activated. Filtration occurs whenever the pump is on, and this often is controlled by a timer with a manual override.

The solar loop is usually added between the filter and the conventional heater, as shown in figure 8–1. A key element in the system is the flow diverter valve. When the valve is in one position, the solar loop is bypassed (when there is inadequate or zero solar gain, or no need for heat). When the valve is in the other position, water is allowed to flow up to the collectors, indicating both a need for heat and adequate solar gain. The diverter valve can be either manually or automatically controlled. An automatic diverter valve is connected to a differential controller. The controller, in turn, is connected to two sensors, one at the outlet from the pool and the other at the collector outlet.

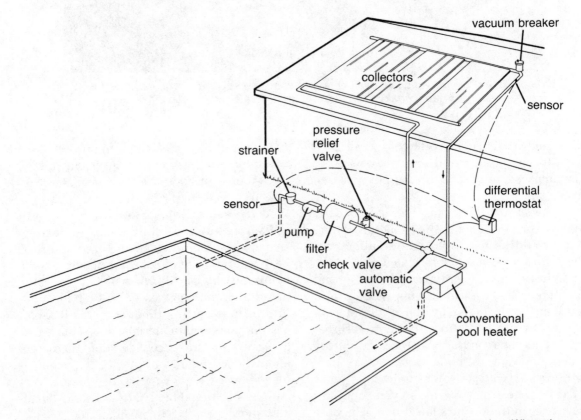

Figure 8–1: The heart of a solar pool heating system is a manual or automatic gate valve. When the valve is closed, water is forced up into the solar loop.

TABLE 8–2

Estimated Collector Size Required to Heat Pool
(as a percentage of pool surface area)

| City | Pool Heating Season | | | |
	June to August	May to September	April to October	All Year
Yuma, Ariz.				
No cover	0	0	20	•
With cover	0	0	0	•
Atlanta, Ga.				
No cover	0	60	•	•
With cover	0	0	80	•
Washington, D.C.				
No cover	0	100	•	•
With cover	0	0	•	•
Los Angeles, Calif.				
No cover	40	70	•	•
With cover	0	0	20	•
Sacramento, Calif.				
No cover	40	70	•	•
With cover	0	0	60	•
Denver, Colo.				
No cover	100	•	•	•
With cover	0	80	•	•
Milwaukee, Wis.				
No cover	100	•	•	•
With cover	20	•	•	•

• Collector area of more than 100% of pool area required.

SOURCE: Harry W. Sigworth, Jr., "Keeping Swimming Pools Warm," unpublished monograph, 1981.

NOTE: The figures given assume that collectors are south facing at a 20-degree tilt; the water is maintained at 80°F during the day; the pool is in a sunny but sheltered site; the pool is protected by an insulating foam cover, 0.1 inch thick; and the cover is removed from 8 A.M. to 6 P.M. daily during the pool heating season.

When there is enough heat in the collectors, the controller powers the diverter valve into the solar collection mode.

The same thing can be done with a common gate valve. When the collectors are higher than the pool, the gate valve is placed between the inlet and outlet of the solar loop, the same as the diverter valve. When the gate valve is open, the water will not run up to the collectors, but will instead follow the path of least resistance straight from the filter to the conventional heater. Close the valve, and water will flow up to the collectors. The gate valve, then, is basically another form of the manual diverter valve. If the collectors are installed below the pool, a second valve is installed on the collector inlet to prevent water from continually, and often needlessly, flow-

Photos 8–5, 8–6: These two pool installations show the flexibility of system design. Collectors can be placed in the best spot for sunshine and good appearance.

ing into the solar loop. This valve can also be manual or automatic. The obvious advantage of using one or two automatic valves is the convenience of not having to attend to a multitude of valve openings and closings. The differential thermostat does the work for you—for a price, of course.

A thermostat and two sensors cost about $250. A system with two automatic valves (needed when the collectors are below the pool) costs about $300. Also needed is a check valve and a vacuum breaker, which cost around $50 for the pair. The same package with manual valves costs about $200. The price of automation is certainly higher, but not by much, and when you look at the cost-effectiveness of the whole system, an extra $50 to $100 doesn't change the payback very much.

There hasn't been any mention of freeze protection in this water-based system because there isn't any need for it. You're not likely to be swimming with Jack Frost anyway, but no matter: Collectors that are above the pool are self-draining, though collectors that are placed below the pool would have to be drained when temperatures approach freezing. (A rubbery SolaRoll collector, by the way, would stretch, not burst, if it froze.)

In planning your system, there are just three sizing requirements to consider. The rule of thumb for collector sizing in cold climates is that collector area should equal about half the surface area of the pool, but this is a highly variable guideline, as table 8–2 shows. Harry Sigworth developed this table as a sizing guideline for the seven cities included in his study. The pipe size for the solar loop should be the same as that used in the existing filter/heater loop. For the pump it's usually possible to use the pump that's already in the filter/heater system, typically a

½ to 1-horsepower unit. If there turns out to be a great deal of added vertical head because the collectors are far from the pool, a second booster pump can be added to the solar loop. Consult the chart in figure 6–5 to see if your pump's horsepower can handle the added feet of head. You want to create a flow rate of around 0.04 gpm per square foot of collector, which is twice the recommended rate for domestic water heating.

When you're looking for a place to put anywhere from 100 to 800 or more square feet of collector, a south-facing roof pitch is an obvious place to investigate. A privacy fence could also be used to support a row of collectors. A south-facing bank below the pool has also been used. Studies have shown, however, that since the tilt angle for pool collectors is relatively shallow, they can be faced west with no significant loss of performance compared to the same collectors facing true south. So you aren't necessarily excluded from having a solar pool if you have a clear western exposure instead of a south-facing location. If space is your problem, you can use less collector area and simply get less of a solar heating benefit. But since solar pool heating is so cost-effective, using even half the recommended area will still give you a good return on your investment.

Different brands of pool absorber will have different installation methods. The tilt angle is different than for domestic water heating collectors because pool collectors are mostly used during warmer weather when the sun is higher in the sky. Thus, a tilt angle of 15 to 20 degrees is recommended. If you're planning to solar heat an indoor pool year-round, tilt the collectors to your latitude angle.

What makes solar pool heating worth the effort? Harry Sigworth's numbers in table 8–1 give a lot of the answers. He also made some calculations on what would be justifiable costs per square foot for an installed system based on current fuel cost (gas, oil, electricity), projected solar savings, projected fuel cost increases (15 percent per year) and the after-tax interest rate on borrowed money. With all those factors involved, the chart in figure 8–2 is mercifully simple to use. (Remember, these results are based on a *model* swimming pool, and they should be used as a general guide to what you can expect a solar system to do for you. Experienced pool installers should be able to provide specific cost

and performance benefit numbers.) From the horizontal axis of the graph find your current energy cost (gas, oil or electricity). From that point, go vertically up to one of the lines that represents the projected savings that a solar system will give you (in terms of therms per square foot per year). Table 8–3 shows the energy savings that can be expected with a solar heater in selected regions. Refer to the table to see which location is closest to your site. If, for example, the savings in your region would be 0.9 therms per square foot per year, go up on the graph to the 1-therm line.

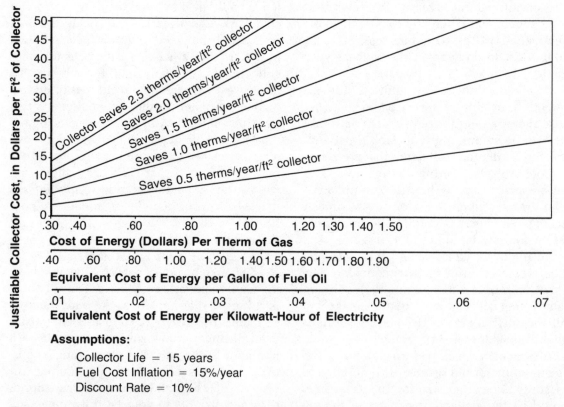

Figure 8–2: The cost of heating a pool with conventional energy sources can be compared with solar pool heating by using this graph. The graph will also help you figure out the justifiable cost per square foot of collector area. The ever-rising prices of oil, gas and electricity make solar pool heating a highly cost-effective investment.

TABLE 8–3

Pool Heating Fuel Savings
Per Square Foot of Unglazed Collector (therms/yr./ft² of collector)

| City | Pool Heating Season | | | |
	June to August	May to September	April to October	All Year
Yuma, Ariz.				
No cover	•	•	0.5	3.0
With cover	•	•	•	1.9
Atlanta, Ga.				
No cover	•	1	1.7	2.4
With cover	•	•	0.6	0.9
Washington, D.C.				
No cover	•	1.2	1.7	2.2
With cover	•	•	1.2	0.8
Los Angeles, Calif.				
No cover	0.7	1.3	1.9	3.3
With cover	•	•	0.9	1.9
Sacramento, Calif.				
No cover	0.8	1.6	2.3	2.8
With cover	•	0.5	0.8	1.4
Denver, Colo.				
No cover	1.3	2.2	2.6	3.3
With cover	•	0.6	1.2	1.4
Milwaukee, Wis.				
No cover	1.5	1.8	2.0	2.5
With cover	•	0.7	1.0	1.3

• Less than 0.5 therms/yr./ft² of collector.

SOURCE: Harry W. Sigworth, Jr., "Keeping Swimming Pools Warm," unpublished monograph, 1981.

NOTE: The figures given assume that the collector is south facing at a 20-degree tilt; the water is maintained at 80°F during the day; the pool is in a sunny but sheltered site; the pool is protected by an insulating foam cover, 0.1 inch thick; and the cover is removed from 8 A.M. to 6 P.M. daily during the pool heating season.

Then go left in a horizontal line to the vertical axis. This axis gives the justifiable system cost in dollars per square foot.

The bottom line is that even without a fat federal tax credit, solar pool heating is a great investment, even after you put in a cover. It gives you a warmer pool and a longer swimming season, and it gives your existing pool heater an all-but-permanent vacation.

Hot Tubs and Spas

The differences between hot tubs and spas (mostly shape and what they're made of) are slight, but they have one thing in common: They need a lot of hot water. When swimming pools get heated up to 80° to 85°F, bathers love it. But people use hot tubs and spas at 105° to 110°F to get their full therapeutic effect, and that requires somewhat more energy for every gallon of water that must be heated.

Hot tubs and spas, like swimming pools, are usually bought with filter/heater packages that are run by a pump that may also power the water jets (a.k.a. Jacuzzi) that give your body a bubbly massage. Also like pools, hot tubs and spas need covers, and because they're somewhat smaller than pools the covers can be as simple as rigid foam insulation (Styrofoam) that's cut to fit. That's an effective cover because its insulation value (R-5 per inch) is somewhat higher than that of a pool cover. In fact, you could float 2 inches of rigid

foam on the water and have a super-insulated pool cover.

The relative smallness and higher operating temperature of hot tubs and spas also warrant the use of insulation around the container. If a deck surrounds a hot tub or spa, the unexposed walls and bottom could be insulated to minimize conductive heat loss. Spas that are being built into the ground should be insulated with rigid foam (sprayed urethane or extruded polystyrene) before the walls and floor are poured.

Solar heating systems for hot tubs and spas are somewhat like those for pools. The

Photo 8–7: Your hot tub or spa can be an ideal retreat that doesn't have to take up much space. The installation shown here (solar collectors in the background) is all built on top of a small, flat-roofed house.

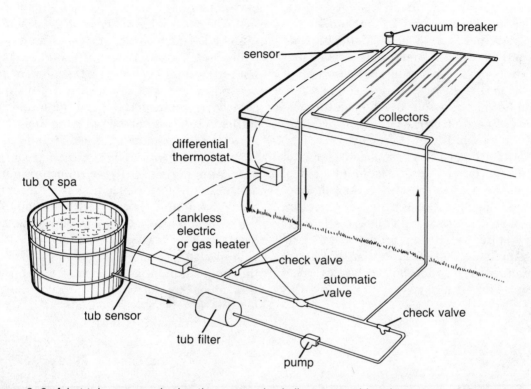

Figure 8–3: A hot tub or spa solar heating system is similar to a pool heating system, with two main exceptions: The collectors have different sizing requirements, and the differential controller in a hot tub or spa system has special temperature-limiting circuitry.

solar collector area should equal 300 percent of the surface area of the hot tub or spa. This means, for example, that a tub with 25 square feet of surface area should be serviced with 75 square feet of collector area.

Two types of collectors can be used. If you want a solar heating system that will operate only in the spring, summer and fall, buy standard unglazed swimming pool heating panels. Since they lack glazing, however, they won't work very well in the cold months. A year-round system requires glazed flat plates, although some contractors recommend glazed

collectors for any installation in all but the warmest climates. The reason is that the higher temperature required for a spa or hot tub mandates the use of a glazed collector. (If you decide to use glazed flat plate collectors, refer to figure 3–15 to determine whether your climate requires one or two layers of glazing.) Freeze protection won't be a concern, because, as in the pool system, the collectors and exposed plumbing will drain whenever the pump is off.

Hot tubs and spas, like swimming pools, act as storage tanks for the heat produced by the collectors. You want the water to be quite warm, but not too warm. To that end, all hot tubs and spas should have automatic, not manual, controls. If the system wasn't limited in its output, water temperatures could climb dangerously high, perhaps up to 150°F. If you dozed off in your overheated tub, you could get cooked. Automatic controls eliminate the possibility of overheating.

A typical solar heating package for a hot tub or spa is similar to a swimming pool package except that the differential thermostat has what is known as a *high-temperature limit switch*. This special circuit keeps water temperatures below approximately 107°F by stopping the pump. This switch should *not* be regarded as an option, but a necessary safety feature. You must have one.

Installation

A simple solar water heating system is plumbed to a hot tub or spa as shown in figure 8–3. The solar loop "tees off" from the hot tub's plumbing between the filter and the conventional heater. An electrically operated gate valve, connected to the differential controller, sends the water up into the solar loop, which uses the same diameter tubing used in the existing loop. The controller's sensors are commonly placed at the collector's outlet and in the drainpipe leading away from the tub. A vacuum breaker must be installed at the highest point in the solar loop so the collectors will drain when the pump is off. There should also be a check valve somewhere in the tub's plumbing to prevent water from flowing backward. For safety's sake, a registered electrician should install the differential controller and its wiring.

9

WORKING WITH A SOLAR CONTRACTOR

Homeowners who don't have time to build and install their own collector systems can rely on the services of a good contractor. There are many different brands of commercially available solar water heaters. You should probably limit your selection to those brands that are available in your area. A good first step is to check the Yellow Pages under "Solar," where you'll probably find the names of several solar contractors. Some dealer/installers have been around for years, and others may be new at it, or they may be plumbers who do a little solar work on the side.

Naturally you want to deal with a good contractor. If there is a nonprofit solar energy association in your area, you may be able to get recommendations or at least an idea of who's been doing what, and for how long. A contractor should be willing to come to your home and provide a free estimate of the cost of the total installation. Some contractors are also willing to estimate, in writing, the number of Btu's or the solar fraction the system will produce in an average year at your property, but if that information isn't forthcoming, you may have to press for it. Last, get the names of several homeowners who own systems installed by the contractors who give you bids.

Talk with these people. Are they satisfied with their systems and with the work of the contractor? If they had it to do over again,

would they do anything differently? If you're interested in doing some of the work yourself, maybe you can find out if the installer is amenable to bidding a partial installation. If you are certain to be hiring out the installation, this book should help you in specifying the work you want done so that every bid you get is based on the same job.

It's important to take the time needed to ask lots of questions of both the contractors and their customers. There's no need to rush into a decision, no matter how excited you may be about the prospect of solar hot water. A solar water heater should not be an impulse buy, but a deliberate and thoroughly considered investment.

Get at least three bids on similar systems. In other words, solicit at least three bids on a batch system, or three bids on a draindown system, or three bids on an antifreeze system, all specifying the same collector area. This way you won't be "mixing apples with oranges," as contractors say. You may have some difficulty finding contractors who deal in batch and drainback systems, as the active antifreeze and draindown systems tend to dominate the contractor scene.

Ask to see a manual explaining the system each contractor wants to sell you. A good system should have a good owner's manual. Finally, before committing yourself to a particular contractor, contact the Better Business Bureau in your area to make sure that the firm is reputable.

Performance Guarantees and Payment Schedules

Lewis Morris, a staff member of the Federal Trade Commission's Bureau of Consumer Protection, stated, "Most of the complaints the government has received concerning solar energy contractors involve performance promises. For example, a contractor might promise a 60 to 80 percent savings in water heating costs, when the actual solar fraction is only 15 to 20 percent. We have also received complaints concerning the quality of manufactured products. Another problem is that installers sometimes are not qualified to do the job. The problem of installation is a big one." He noted that in most states solar contractors do not have to be certified.

These problems have prompted Morris to offer consumers the following advice: First, research the solar equipment you're thinking about buying; read as much literature as possible (Morris recommends *Consumer Reports* magazine). Second, research the contractor you're thinking about hiring, and always ask for a list of clients. Third, demand a written performance guarantee as part of the contract

Photo 9–1: A Btu meter records the exact amount of heat produced by the system.

with the installer. "The contract might guarantee, for example, a 50 percent solar fraction, which equates to X number of Btu's per year, which equates to X amount of water heating savings a year," Morris said. "The contract should also specify compensation or removal of the system and a cash refund if the performance guarantee is not met." (You can use the heat gain calculation in chapter 3 to estimate the number of Btu's your system should produce. This, however, will only give you an estimate. For an exact measurement you could use a device called a *Btu meter*, like the one shown in photo 9–1. It is installed in the collector outlet and totalizes the Btu output of the system. To check flow rates you can buy or borrow—since you really only need to check flow rate once—a *flow meter*, like the one in photo 9–2.)

Morris pointed out that claimed system performance is one of the major reasons why a homeowner might decide to buy a particular solar water heater, so it is only fair that the contractor be willing to guarantee it in writing. "If a contractor is unwilling to provide a written performance guarantee," he said, "the consumer would be best advised to find another contractor."

Tom Super, of the Washington, D.C.-based National Association of Solar Contractors, disagrees. "I certainly would not recommend to contractors that they guarantee system performance," he said. "There are just too many variables, such as hot water use and weather. Not long ago a consumer complained to us that his electric bill had not decreased after he bought a solar water heater. We looked into it and found that was because he was using more hot water, and drawing more auxiliary power. That's certainly not the contractor's fault."

Super recommends a delayed payment schedule as the best protection for con-

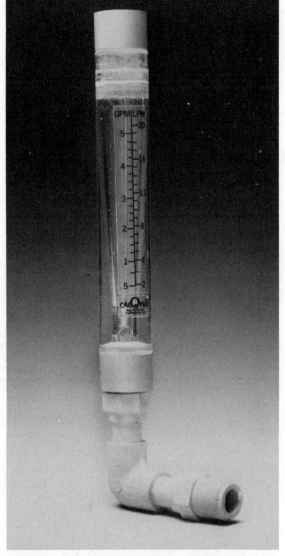

Photo 9–2: A flow meter tells you how fast the collector fluid is flowing. Some pumps are so quiet that a flow meter is the only way to know for sure.

sumers. "The contract you sign with an installer should specify a 40 percent down payment at the start of the job, another 40 percent when all the equipment has been installed, and the remaining 20 percent only after the system has been tested and filled," he said. (Super also said that he thought it was only a matter of time before solar installers are certified in all states. "We are either going to certify ourselves, or someone will do it for us," he said.) As it turns out, though, there has been a great deal of work on the certification of commercially available systems, if not the contractors who install them. The certification procedures involve scientifically controlled tests that really do show how one manufacturer's collector does against others. The Solar Rating and Certification Corporation (SRCC) was formed to develop a rating and certification system that could be applied nationally. The system uses a standard method of testing and comparison (95–81 SPR) to come up with a Solar Performance Ratio (SPR). The American Society of Heating, Refrigeration and Air-Conditioning Engineers (ASHRAE) has also developed a method (ASHRAE 95–81) toward the same ends. You can learn more about the results of tests done by these methods by reading through reports that are listed in Further Reading, at the end of the book. You can also get direct information by calling your state energy office and inquiring about ratings and certifications for different brand names.

Common Installation Errors

In recent years investigators funded by the U.S. Department of Energy and the solar industry have traveled around the country scrutinizing professionally installed systems. The investigators found many systems that were not properly installed, and there were certain mistakes that were often repeated.

One study was conducted by the Northeast Solar Energy Center (a former Department of Energy contractor) and the National Association of Solar Contractors. In this 1981 study six teams of inspectors trekked across New England, Pennsylvania, New York and New Jersey and made on-site evaluations of 170 systems. The investigators concluded the study by recommending certification for solar contractors.

Here, in abbreviated form (a more thorough report of this study can be found in the October 1981 issue of *Solar Age* magazine), are some of the findings:

• Collectors were often incorrectly oriented and tilted.

• Collectors were installed in the shade. Thirty-five percent of the systems studied had collectors that were at least partially shaded. Most often trees were the culprits. Occasionally collectors that had been installed in a sawtooth fashion to some extent shaded each other.

• Collectors were inadequately attached to roofs. The inspectors found collectors that were attached only to the roof sheathing. Roof-mounted collectors should always be securely fastened to the roof rafters.

• Collectors were sometimes too heavy for the roof. Nineteen homes were found to have roofs that were not strong enough to carry the weight of the collectors.

• Untreated wood was used. The investigators preferred collector racks that were fabricated from similar metals. If you prefer to use wood it should be pressure-treated pine, redwood or cedar.

• Collector glazings were obscured by internal condensation and outgassing. Urethane installation inside the collectors can sometimes generate gases that fog the glazing and reduce the transmittance of the glazing. Condensation can also impair the thermal properties of the insulation.

• Metal components were corroded. Screws, nuts, bolts and fasteners holding collector panels together were occasionally rusted.

• Sensors were installed improperly.

• Piping and roof penetrations were sometimes poor. The collectors in one-fifth of the systems were undrainable. Some of the piping was not installed to "accepted pipe hanging standards." Seven of the eight systems that used CPVC tubing were found to have "unacceptable distortion and waving in the loop." One in five of the systems had leaky roof penetrations. A snug-fitting, shoe-type collar should be used to prevent leakage into the roof.

• Pipe runs were inadequately insulated. There were three problems: low R-value, poor seals and ultraviolet degradation. All solar loop tubing should be insulated to at least R-4; tubing that has a diameter greater than 1 inch should be insulated to at least R-6. The seams in wrap-around insulation should be sealed with contact cement. Uncovered elastomer insulation was prone to ultraviolet degradation. Some contractors were found to have sealed the outdoor insulation with ordinary house paint, which quickly degraded. The inspectors recommended sealing the pipe insulation in metal cladding or PVC jackets.

• Heat transfer fluids were too acidic or watery. Glycol solutions should be checked from time to time to make sure they have not turned acidic. Many of the systems were filled with antifreeze that contained too much water. It was theorized that several installers, after pressure-testing the solar loop with water, did not drain all the water out of the loop before adding the antifreeze solution.

• Necessary valves and vents were sometimes missing. Storage tanks should always have a pressure-and-temperature re-

lief valve to prevent dangerous pressure buildups. The solar loop of a closed-loop system should also have a pressure relief valve. Mixing valves should be added to the hot water outlet pipe of the storage tank to eliminate the possibility of scalds. Air vents should be added to the highest point of the solar loop so that the collectors can drain. The air vents should be installed following the manufacturer's instructions.

• Some storage tanks were not properly insulated. All storage tanks should be wrapped in a blanket of insulation (6-inch fiberglass), as described in chapter 1. The tanks should also be installed in a manner that will make them easily accessible for service.

• Owner's manuals were not given to homeowners. "In nearly one-third of the cases studied, a system manual was not left with the homeowner," the report stated. "Several homeowners admitted to a total lack of knowledge about their systems."

Other mishaps with active systems have included malfunction and improper installation of the differential thermostat. Make sure your installer does a complete check-out of the control system after it's installed, with you standing there to see how it all works. It has also happened that power outages have done funny things to system controls, so if an outage should occur, it's a good idea to check your control system to make sure that all is well.

Another enlightening study was conducted by the National Solar Data Network. Since 1977 the organization has collected data from over 150 systems. An article in *Solar Age* (February, 1982) boiled the findings down to three directives: "Keep it simple, keep it cool (the collector) and keep it cheap." To that end, a simple system is often to be preferred over a more complex one. "A solar heating or cooling installation is a fairly complex project to begin with," wrote C. J. Kelly in *Solar Age*. "It almost always contains a conventional system as a backup. . . . As the complexity increases, the number and size of the pumps, fans, motorized valves and dampers also grows. I have seen very complex systems use half as much electrical energy as the solar energy delivered."

Solar heating systems are most efficient when operating at the lowest possible temperatures. Low temperatures prevent excessive "standby losses," or heat loss through pipe runs and storage tanks. Collectors also operate most efficiently at low temperatures.

It's easy to understand why a solar water heating system should be "cheap," or inexpensive. But you'd be amazed at the number of homeowners who would buy a $4000 antifreeze system when a $1000 batch heater would work just as well.

Be the Boss

Don't be afraid to scrutinize your contractor's work. If you notice a mistake, speak up. It will be easier to make corrections while the job is still in progress.

If you want to do the work yourself, you may be able to find a contractor who'll plan your system for a fee and sell you the equipment. (This may not be a good idea for some solar components, as the manufacturer's warranty may be valid only if the component is installed by the contractor.)

This raises another possibility for saving money. A contractor is essentially a middle-man who buys supplies from a wholesaler and sells to you at retail, but you may find a contractor who'll be willing to use materials that you supply. Of course, this will only work if you can buy materials for a lower price than what the contractor would charge you.

While the contractor might agree to use some of your materials, he'll probably insist on installing his own solar components. It's easy to see why. Like all entrepreneurs, solar contractors are in business to make money. Some of the contractor's money is made by reselling equipment to you at retail. But most of the money is usually made from the installation. Consequently, the contractor selects systems that are best for his business as well as for his customers. The selection of the solar components is an important business decision, and you may have a hard time finding a contractor who's willing to install another type of system.

Don't Forget Conservation

These days you should expect to pay between $2500 and $4500 for an installed solar water heating system. The best way to reduce this cost is to follow the conservation measures outlined in chapter 1. Some contractors may not be proponents of conservation, perhaps out of ignorance, perhaps out of a desire to sell you as much collector area as they can. If you reduce the amount of hot water you need, you'll need a smaller, and less expensive, solar system. It is fitting that this book ends with the same thought with which it began: conservation.

APPENDIX 1
SOURCES FOR HARDWARE

Hardware Catalogs

Peoples Solar Sourcebook
Solar Usage Now, Inc.
P.O. Box 306
420 E. Tiffin St.
Bascom, OH 44809
(800) 537-0985
(419) 937-2226

Solar Components Corporation's
  Solar Catalog
Solar Components Corp.
P.O. Box 237
Manchester, NH 03105
(603) 668-8186

Chapter I

Greywater Recycling Systems

The Burning Log
P.O. Box 438
Lebanon, NH 03766
(603) 448-4360

or

P.O. Box 8519
Aspen, CO 81612
(303) 925-8968

**Low-Flow Shower Heads
and Faucet Aerators**

(Local hardware stores)

Pipe Insulation

Armstrong World Industries, Inc.
Box 3001
Lancaster, PA 17604
(717) 397-0611

Celotex Corp.
1500 N. Dale Mabry Hwy.
Tampa, FL 33607
(813) 871-4133

CPR
Div. of Upjohn
555 Alaska Ave.
Torrance, CA 90503
(213) 320-3550

Knauf Fiber Glass
240 Elizabeth St.
Shelbyville, IN 46176
(317) 398-4434

Northeast Specialty Insulations, Inc.
1 Watson Pl.
Saxonville, MA 01701
(617) 877-0721

Pipe Insulation — *continued*

Owens-Corning Fiberglas Corp.
Mechanical Div.
Fiberglas Tower, OH 43659
(419) 248-8102

Pittsburgh Corning Corp.
800 Presque Isle Dr.
Pittsburgh, PA 15239
(412) 327-6100

Sentinel Energy Saving Products, Inc.
Div. of Packaging Industries Group
130 North St.
Hyannis, MA 02601
(617) 775-5220

Solar Components Corp. (distributor)
P.O. Box 237
Manchester, NH 03105
(603) 668-8186

Teledyne Mono-Thane
1460 Industrial Pkwy.
Akron, OH 44310
(216) 633-6100

Thermacor Process, Inc.
P.O. Box 4529
500 N.E. 23rd St.
Ft. Worth, TX 76106
(817) 624-1181

Voltek Inc.
Lawrence, MA 01843
(800) 225-0668

Preinsulated Pipes

Insultek Corp.
P.O. Box 329
Warwick, NY 10990
(914) 986-5040

Rovanco Corp.
I–55 and Frontage Rd.
Joliet, IL 60436
(815) 741-6700

Thermacor Process, Inc.
P.O. Box 4529
500 N.E. 23rd St.
Ft. Worth, TX 76106
(817) 624-1181

Stack Damper for Gas Water Heaters

American Metal Products Co.
Div. of Masco Corp.
6100 Bandini Blvd.
Los Angeles, CA 90040
(213) 726-1941
(Ameri-Therm brand)

Solar Usage Now, Inc. (distributor)
P.O. Box 306
420 E. Tiffin St.
Bascom, OH 44809
(800) 537-0985
(419) 937-2226

(Check with local plumbing retailers or
  wholesalers)

Tankless (Instantaneous) Water Heaters

Chronomite Laboratories, Inc.
21011 S. Figueroa St.
Carson, CA 90745
(213) 533-0409
(electric)

John Condon Co., Inc. (distributor)
1103 N. 36th St.
Seattle, WA 98103
(206) 632-5600
(gas)

In-Sink-Erator Div.
Emerson Electric Co.
4700 21st St.
Racine, WI 53406
(414) 554-5432
(electric)

International Technology Sales Corp.
  (distributor)
7344–G S. Alton Way
Englewood, CO 80112
(303) 771-6160
(electric and gas)

Little Giant Manufacturing Co., Inc.
P.O. Box 518
907 7th St.
Orange, TX 77630
(713) 883-4246
(electric and gas)

North American Solar Development Corp.
P.O. Box 65
Cabin John, MD 20818
(703) 241-8886
(electric)

Paloma Industries, Inc.
241 James St.
Bensenville, IL 60106
(312) 595-8778
(gas)

Pressure Cleaning Systems, Inc. (distributor)
Junkers Heaters
612 N. 16th Ave.
Yakima, WA 98902
(509) 452-6607
(gas)

The Tankless Heater Corp.
Melrose Square
Greenwich, CT 06830
(203) 661-2102
(electric and gas)

Timers for Electric Water Heaters

Solar Components Corp. (distributor)
P.O. Box 237
Manchester, NH 03105
(603) 668-8186

(Local hardware and electrical supply stores)

Timers for Gas Water Heaters

Minnesota NRG Co.
Div. of S.R. Sikes Co.
200 N. 3rd St.
Minneapolis, MN 55401
(612) 333-6271 ·

Chapter 4

Black Graphite Paint

Solar Usage Now, Inc. (distributor)
P.O. Box 306
420 E. Tiffin St.
Bascom, OH 44809
(800) 537-0985
(419) 937-2226

Commercial Batch Collectors

Cornell Energy, Inc.
245 S. Plumer Ave.
Suite 27
Tucson, AZ 85719
(602) 882-4060

Environment|One Corp.
2773 Balltown Rd.
Schenectady, NY 12309
(518) 346-6161

Gulf Thermal Corp.
P.O. Box 1273
Sarasota, FL 33578
(813) 957-0106

Commercial Batch Collectors — *continued*

Hemet Solar, Inc.
26740 Cawston Ave.
P.O. Box 1709
Hemet, CA 92343
(714) 652-2891

Nature's Way Energy Systems
P.O. Rte. 8
Old Homestead Hwy. (Rte. 32)
Keene, NH 03431
(603) 352-1186

Solar Components Corp. (distributor)
P.O. Box 237
Manchester, NH 03105
(603) 668-8186

Zomeworks
P.O. Box 25805
1221 Edith Blvd., N.E.
Albuquerque, NM 87125
(505) 242-5354

Glazing

AFG Industries, Inc.
1400 Lincoln St.
Kingsport, TN 37662
(800) 251-0441
(615) 245-0211
(glass)

American Durafilm Co., Inc. (distributor)
2300 Washington St.
Newton Lower Falls, MA 02162
(617) 969-9800
(Teflon)

Cadillac Plastic and Chemical Co.
   (distributor)
1221 Bowers St.
Birmingham, MI 48012
(313) 646-5100
(Lucite L, Lucite SAR and Teflon)

CalWest Energy Services, Inc. (distributor)
2672 Via Pacheco
Palos Verdes Est., CA 90274
(213) 541-7441
(Tedlar)

Commercial Plastics and Supply Corp.
   (distributor)
98–31 Jamaica Ave.
Richmond Hill, NY 11418
(212) 441-1500
(Lucite L and Lucite SAR)

CYRO Industries
P.O. Box 488
155 Tice Blvd.
Woodcliff Lakes, NJ 07675
(201) 930-0100
(ACRYLITE GP and EXOLITE Acrylic)

Filon Div. of Vistron Corp.
12333 S. Van Ness Ave.
Hawthorne, CA 90250
(213) 757-5141
(Solar-E)

General Electric Co.
Plastics Div.
Specialty Plastics Dept.
Sheet Products Section
1 Plastics Ave.
Pittsfield, MA 01201
(413) 494-1110
(Lexan)

General Glass International Corp.
542 Main St.
New Rochelle, NY 10801
(914) 235-5900
(glass)

Glasteel, Inc.
Highway 57E
Collierville, TN 38057
(800) 238-5546
(glass)

Hordis Brothers, Inc.
825 Hylton Rd.
Pennsauken, NJ 08110
(609) 662-0400
(glass)

Lasco Industries
3255 E. Miraloma Ave.
Anaheim, CA 92806
(714) 993-1220
(Crystalite-T)

Martin Processing, Inc.
Film Div.
P.O. Box 5068
Martinsville, VA 24112
(703) 629-1711
(LLumar)

Rohm & Haas Co.
Independence Mall West
Philadelphia, PA 19105
(215) 592-3460
(Plexiglas G, Tuffak A and Tuffak-Twinwal)

Sheffield Plastics, Inc.
P.O. Box 248
Salisbury Rd.
Sheffield, MA 01257
(413) 229-8711
(Acry-Pane and Poly-Glaz)

Solar Components Corp. (distributor)
P.O. Box 237
Manchester, NH 03105
(603) 668-8186
(Sun-Lite Premium II)

Solar Research Div. (distributor)
Refrigeration Research, Inc.
525 N. 5th St.
Brighton, MI 48116
(313) 227-1151
(Tedlar)

Structured Sheets, Inc.
196 E. Camp Ave.
Merrick, NY 11566
(516) 546-4868
(Qualex)

3M Co.
Energy Control Products
3M Center
Building 220–8E
St. Paul, MN 55144
(612) 736-2388
(Flexigard 7410 and 7415)

3M Co.
Transparent Insulation Products
Industrial and Consumer Sector/3M
53–03–01 3M Center
St. Paul, MN 55144
(800) 328-1300
(SunGain)

Heating Cable (heat tape)

Raychem Corp.
Consumer/Commercial Div.
P.O. Box 8036
Redwood City, CA 94063
(415) 361-3333

**Reflective Polyester Film
(e.g. Aluminized Mylar)**

Chemplast, Inc. (distributor)
150 Dey Rd.
Wayne, NJ 07470
(201) 696-4700

Solar Usage Now, Inc. (distributor)
P.O. Box 306
420 E. Tiffin St.
Bascom, OH 44809
(800) 537-0985
(419) 937-2226

Selective Coating Foil for Tank

Berry Solar Products
2850 Woodbridge Ave.
Edison, NJ 08837
(201) 549-0700

Chapter 5

Absorber Plates

Approtech Solar Products
770 Chestnut St.
San Jose, CA 95110
(408) 297-6527
(aluminum and copper absorber plate)

Bio-Energy Systems, Inc.
221 Canal St.
Ellenville, NY 12428
(914) 647-6700
(EPDM absorber plate)

Phelps Dodge Solar Enterprises
1590 S. Sinclair St.
Anaheim, CA 92806
(714) 978-1122
(all-copper absorber plate)

Shelley Radiant Ceiling Co., Inc.
456 W. Frontage Rd.
Northfield, IL 60093
(312) 446-2800
(copper and aluminum absorber plate)

Solar Development, Inc.
Garden Industrial Park
3630 Reese Ave.
Riviera Beach, FL 33404
(305) 842-8935
(all-copper absorber plate)

Sunplate, Inc.
Subsidiary of Columbia Chase Corp.
Solar Energy Div.
55 High St.
Holbrook, MA 02343
(617) 767-0513
(all-copper absorber plate)

Terra-Light
Div. of Butler Manufacturing Co.
54 Cherry Hill Dr.
Danvers, MA 01923
(617) 663-2075
(all-copper absorber plate)

Thermatool Corp.
P.O. Box 522
280 Fairfield Ave.
Stamford, CT 06902
(203) 357-1555
(all-copper absorber plate)

Western Solar Development, Inc.
1236 Callen St.
Vacaville, CA 95688
(707) 446-4411
(all-copper absorber plate)

Collector Racks

Gascoigne Industrial Products, Inc.
Kee Klamp Div.
P.O. Box 207
Buffalo, NY 14225
(716) 685-1250

Miller & Sun Enterprises, Inc.
9450 S.W. Tigard St.
Tigard, OR 97223
(503) 620-2616

Solar-Eye Products, Inc.
1300 N.W. McNab Rd.
Building G & H
Ft. Lauderdale, FL 33309
(305) 974-2500

Sunracks, Div. of Sunsearch, Inc.
P.O. Box 275
Guilford, CT 06437
(203) 453-6591

Sunworks, Inc.
P.O. Box 3900
Somerville, NJ 08876–1270
(201) 469-0399

Tucson Solar Component
  Manufacturing Co.
5741 E. Scarlett St.
Tucson, AZ 85711
(602) 747-3826

Commercial Flat Plate Collectors

Acrosun Industries, Inc.
1800 Shelton Dr.
Hollister, CA 95023
(408) 637-8191

Advance Energy Technologies, Inc.
P.O. Box 387
Clifton Park, NY 12065
(518) 371-2140

Alpha Solarco, Inc.
Suite 2530
1014 Vine St.
Cincinnati, OH 45202
(513) 621-1243

AMAX, Inc.
Climax Molybdenum Co.
AMAX Center
Greenwich, CT 06836
(203) 629-6400

American Solar King Corp.
7200 Imperial Dr.
P.O. Drawer 7399
Waco, TX 76714–7399
(817) 776-3860

Ametek
Power Systems Div.
1025 Polinski Rd.
Ivyland, PA 18974
(215) 441-8770

Approtech Solar Products
770 Chestnut St.
San Jose, CA 95110
(408) 297-6527

ARCO Comfort Products Co.
302 Nichols Dr.
Hutchins, TX 75141
(214) 225-7351

Commercial Flat Plate Collectors — *continued*

Bio-Energy Systems, Inc.
221 Canal St.
Ellenville, NY 12428
(914) 647-6700

Colt, Inc.
1750 Commerce Way
Paso Robles, CA 93446
(805) 238-9400

Columbia Chase Solar Energy Div.
55 High St.
Holbrook, MA 02343
(617) 767-0513

Daystar Corp.
P.O. Box 1160
90 Cambridge St.
Burlington, MA 01803
(617) 272-8460

Dell Solar Industries, Inc.
1 Second St.
New Rochelle, NY 10801
(914) 235-4141

Dixon Energy Products, Inc.
1102 Center St.
Ludlow, MA 01056
(413) 589-0147

Electra Israel Ltd.
Div. of Airlex Industries
440 Park Ave. S.
New York, NY 10016
(212) 683-1460

Energy Design Corp.
1756 Thomas Rd.
P.O. Box 34160
Memphis, TN 38134
(901) 382-3000

Energy Transfer Systems, Inc.
8505 Sunstate St.
Tampa, FL 33614
(813) 885-7657

Grumman Allied Industries, Inc.
445 Broad Hollow Rd.
Plant 9, Floor #4
Melville, NY 11747
(516) 454-8600

Gulf Thermal Corp.
P.O. Box 1273
Sarasota, FL 33578
(813) 957-0106

Halstead and Mitchell
Div. of Halstead Industries, Inc.
P.O. Box 1110
Scottsboro, AL 35768
(205) 259-1212

Heliodyne, Inc.
700 S. 4th St.
Richmond, CA 94804
(415) 237-9614

KTA Solar, Inc.
13856 A Park Center Rd.
Herndon, VA 22071
(703) 471-5405

Langley Products, Inc.
Four Pines Bridge Rd.
Beacon Falls, CT 06403–1399
(203) 888-0534

Lennox Industries, Inc.
P.O. Box 400450
Dallas, TX 75240
(214) 783-5000

Miller & Sun Enterprises, Inc.
9450 S.W. Tigard St.
Tigard, OR 97223
(503) 620-2616

Mor-Flo Industries, Inc.
18450 S. Miles Rd.
Cleveland, OH 44128
(216) 663-7300

National Solar Corp.
5627 N. 52d Ave.
Glendale, AZ 85301
(602) 242-9325

Northern Solar Power Co.
311 S. Elm St.
Moorhead, MN 56560
(218) 233-2515

Novan Energy, Inc.
1630 N. 63rd St.
Boulder, CO 80301
(800) 525-9518
(303) 447-9193

Radco Products, Inc.
2877 Industrial Pkwy.
Santa Maria, CA 93455
(805) 928-1881

Ramada Energy Systems, Inc.
1421 S. McClintock Dr.
Tempe, AZ 85281
(602) 273-4300

Raypak, Inc.
31111 Agoura Rd.
Westlake Village, CA 91361
(213) 889-1500

Revere Solar & Architectural Products, Inc.
P.O. Box 151
Rome, NY 13440
(315) 338-2401

Reynolds Metals Co.
Energy Products Group
P.O. Box 27003
Richmond, VA 23261
(800) 368-3015

Sealed Air Corp.
3433 Arden Rd.
Hayward, CA 94545
(415) 887-7000

Solar Alternative, Inc.
71 Main St.
Brattleboro, VT 05301
(802) 257-4528

Solar Corp. of America (SOLCOA), Inc.
105 N. Hanks St.
Rome, GA 30161
(404) 291-6956

Solar Development, Inc.
3630 Reese Ave.
Riviera Beach, FL 33404
(305) 842-8935

Commercial Flat Plate Collectors — *continued*

Solar Development, Inc.–Northwest
1254 Wilson Ave.
Pocatello, ID 83201
(208) 233-6563

Solar-Eye Products, Inc.
1300 N.W. McNab Rd.
Building G & H
Ft. Lauderdale, FL 33309
(305) 974-2500

Solar Industries, Incorporated
2300 Hwy. 34
Manasquan, NJ 08736
(201) 223-8100

Solar Living, Inc.
P.O. Box 12
Netcong, NJ 07857
(201) 691-8483

Solar One, Inc.
777 Seahawk Cir.
Virginia Beach, VA 23452
(804) 427-6300

Solar Oriented Environmental Systems, Inc.
10639 S.W. 185 Terrace
Miami, FL 33157
(305) 233-0711

Solar Research, Inc.
P.O. Box 869
Brighton, MI 48116
(313) 227-1151

Solar Source, Inc.
Rt. 1, Hwy. 11–E
P.O. Box S.U.N.
Chuckey, TN 37641
(615) 257-6755

Solar Span International, Inc.
285 Queen St.
Southington, CT 06489
(203) 621-6728

Solar Systems by Sun Dance, Inc.
13939 N.W. 60th Ave.
Miami Lakes, FL 33014
(305) 557-2882

Solarsystems Industries Ltd.
#2–11771 Horseshoe Way
Richmond, BC V7A 4S5
Canada
(604) 271-2621

Solartec, Inc.
250 Pennsylvania Ave.
Salem, OH 44460
(216) 332-9100

Solar Unlimited, Inc.
37 Traylor Island
Huntsville, AL 35801
(205) 534-0661

Solar Usage Now, Inc. (distributor)
P.O. Box 306
420 E. Tiffin St.
Bascom, OH 44809
(800) 537-0985
(419) 937-2226

Solergy Co.
7216 Boone Ave. N.
Minneapolis, MN 55428
(612) 535-0305

Southwest Ener-Tech, Inc.
3020 S. Valley View Blvd.
Las Vegas, NV 89102
(702) 876-5444

Sun-Bank, Inc.
924 N. Main St.
Wichita, KS 67203
(316) 265-0866

Sunearth Solar Products Corp.
352–E Godshall Dr.
Harleysville, PA 19438
(215) 256-6648

Sunloop Corp.
833–A N. Market Blvd.
Sacramento, CA 95834
(916) 924-7244

Suntime, Inc.
6 Main Hill
P.O. Box 401
Bridgton, ME 04009
(207) 647-5240

Sunworks, Inc.
P.O. Box 3900
Somerville, NJ 08876–1270
(201) 469-0399

Thomason Solar Home, Inc.
609 Cedar Ave.
Fort Washington, MD 20744
(301) 292-5122

Universal Solar Development, Inc.
1505 Sligh Blvd.
Orlando, FL 32806
(305) 423-8727

U.S. Solar Corp.
P.O. Drawer K
Hampton, FL 32044
(904) 468-1517

Vulcan Solar Industries, Inc.
6 Industrial Dr.
Smithfield, RI 02917
(401) 231-4422

Western Solar Development, Inc.
1236 Callen St.
Vacaville, CA 95688
(707) 446-4411

Zomeworks Corp.
P.O. Box 25805
1221 Edith Blvd., N.E.
Albuquerque, NM 87125
(505) 242-5354

Kits for D-I-Y Collectors

The Back Forty
P.O. Box 157
Buckfield, ME 04220
(207) 336-2021

Miller & Sun Enterprises, Inc.
9450 S.W. Tigard St.
Tigard, OR 97223
(503) 620-2616

Kits for D-I-Y Collectors — *continued*

Solar Living, Inc.
P.O. Box 12
Netcong, NJ 07857
(201) 691-8483

Solar Usage Now, Inc. (distributor)
P.O. Box 306
420 E. Tiffin St.
Bascom, OH 44809
(800) 537-0985
(419) 937-2226

Roof Penetration Kits

Aardvark & Sun Solar, Inc.
P.O. Box 268
South Dennis, MA 02660
(617) 394-6391
(all-copper flashing)

Oatey Co.
4700 W. 160th St.
Cleveland, OH 44135
(800) 321-9532
(galvanized flashing)

Specialty Products Co.
P.O. Box 186
Stanton, CA 90680
(714) 828-9730
(thermoplastic flashing)

Chapter 6

Check Valves for Thermosiphon Systems

General Solar Technology, Inc.
P.O. Box 570502
10890 Quail Roost Dr.
Miami, FL 33157
(305) 255-5320

Zomeworks Corp.
P.O. Box 25805
1221 Edith Blvd., N.E.
Albuquerque, NM 87125
(505) 242-5354

Complete Flat Plate Systems

Con-Serv Corp.
3341 E. Corona Ave.
Phoenix, AZ 85040
(602) 243-3246
(drainback)

Environment | One Corp.
2773 Balltown Rd.
Schenectady, NY 12309
(518) 346-6161
(thermosiphon)

Chris Fried Solar
RD #3
Box 229G
Catawissa, PA 17820
(717) 356-2501
(drainback)

Grumman Allied Industries, Inc.
445 Broad Hollow Rd.
Plant 9, Floor #4
Melville, NY 11747
(516) 454-8600
(antifreeze, draindown, recirculation)

KTA Solar, Inc.
13856–A Park Center Rd.
Herndon, VA 22071
(703) 471-5405
(drainback)

Ottawa Solartronics, Ltd.
1070 Morrison Dr.
Ottawa, ON K2H 8K7
Canada
(613) 820-3274
(drainback)

Paolino Energy Products, Inc.
726 Kaighns Ave.
Camden, NJ 08103
(609) 541-4614
(antifreeze, drainback)

Raypak, Inc.
31111 Agoura Rd.
Westlake Village, CA 91361
(213) 889-1500
(antifreeze, drainback, draindown,
   recirculation)

Revere Solar & Architectural Products, Inc.
P.O. Box 151
Rome, NY 13440
(315) 338-2401
(antifreeze)

Solar Development, Inc.
3630 Reese Ave.
Riviera Beach, FL 33404
(305) 842-8935
(antifreeze, drainback, draindown, recirculation, thermosiphon)

Solar Living, Inc.
P.O. Box 12
Netcong, NJ 07857
(201) 691-8483
(draindown)

Suntime, Inc.
6 Main Hill
P.O. Box 401
Bridgton, ME 04009
(207) 647-5240
(phase change)

Sun-West Solar Systems, Inc.
4024 E. Broadway Rd.
Suite 1001
Phoenix, AZ 85040
(602) 243-6171
(drainback, recirculation)

Trendsetter Industries
10183 Croydon Way
Suite E
Sacramento, CA 95827
(916) 361-0107
(drainback)

Differential Thermostats
for Solar DHW Systems

American Solar Heat Corp.
7 National Pl.
Danbury, CT 06810
(203) 748-5554

Dan-Mar Co., Inc.
RR 2
Box 338B
Wikel Rd.
Huron, OH 44839
(419) 433-4479

Heliotrope General
3733 Kenora Dr.
Spring Valley, CA 92077
(714) 460-3930

H. I. Square
2611 Old Okeechobee Rd.
West Palm Beach, FL 33409
(305) 686-8400

Independent Energy
P.O. Box 860
42 Ladd St.
East Greenwich, RI 02818
(401) 884-6990

JBJ Controls
P.O. Box 1256
Idaho Falls, ID 83402
(208) 522-2200

Differential Thermostats for Solar DHW Systems — *continued*

Johnson Controls, Inc.
2221 Camden Ct.
Oak Brook, IL 60521
(312) 654-4900

Natural Power, Inc.
Francestown Tpk.
New Boston, NH 03070
(603) 487-5512

Photocomm
7745 E. Redfield Rd.
Scottsdale, AZ 85260
(602) 948-8003

Pyramid Controls
421–16 N. Buchanan Cir.
Pacheco, CA 94553
(415) 827-0160

Rho Sigma, Inc.
Sub. of Watsco, Inc.
1800 W. 4th Ave.
Hialeah, FL 33010
(305) 885-1911

Richdel, Inc.
Solar Div.
P.O. Drawer A
Carson City, NV 89701
(702) 882-6786

Robertshaw Controls Co.
Temperature Controls Marketing Group
100 W. Victoria St.
Long Beach, CA 90805
(213) 636-8301

Solarsystems Industries, Ltd.
#2–11771 Horseshoe Way
Richmond, BC V7A 4S5
Canada
(604) 271-2621

Tour & Andersson, Inc.
652 Glenbrook Rd.
Stamford, CT 06906
(203) 324-0106

Wolfway Product Consultants, Inc.
RD 1, Box 1135
Tamaqua, PA 18252
(717) 668-4359

Drainback Modules (preassembled components)

Bio-Energy Systems, Inc.
221 Canal St.
Ellenville, NY 12428
(914) 647-6700

Draindown Valves

Mor-Flo Industries, Inc.
18450 S. Miles Rd.
Cleveland, OH 44128
(216) 663-7300

Richdel, Inc.
Solar Div.
P.O. Drawer A
Carson City, NV 89701
(702) 882-6786

Sunspool Corp.
439 Tasso St.
Palo Alto, CA 94301
(415) 324-2022

Suntron Inc.
7630 Race Rd.
North Ridgeville, OH 44039
(216) 777-7998

Heat Transfer Fluids for Flat Plate Collectors (antifreeze)

Bray Oil Co., Inc.
Div. of Burmah-Castrol, Inc.
2698 White Rd.
Irvine, CA 92714
(714) 850-1151
(synthetic hydrocarbon base)

Camco Manufacturing, Inc.
121 Landmark Dr.
Greensboro, NC 27409
(800) 334-2004
(propylene glycol base)

Dow Chemical U.S.A.
Organic Chemicals Marketing
9008 Building
Midland, MI 48640
(517) 636-1000
(propylene glycol base)

Dow Corning Corp.
P.O. Box 1–M
Midland, MI 48640
(800) 248-2345
(800) 292-2323 (in Michigan)
(silicone liquid)

Exxon Co., U.S.A.
P.O. Box 3351
Houston, TX 77001
(800) 231-6600
(petroleum base)

Nutek, Inc.
Route 85
Amston, CT 06231
(203) 537-2387
(propylene glycol, ethylene glycol and
nonglycol/nonsilicone bases)

Resource Technology Corp.
1 Alcap Ridge
Cromwell, CT 06416
(203) 635-0267
(organic base; nonaqueous)

Solar Alternative, Inc.
71 Main St.
Brattleboro, VT 05301
(802) 257-4528
(propylene glycol base)

Solar Usage Now, Inc. (distributor)
P.O. Box 306
420 E. Tiffin St.
Bascom, OH 44809
(800) 537-0985
(419) 937-2226
(glycol solution; Sun-Temp)

Sunworks
P.O. Box 3900
Somerville, NJ 08876
(201) 469-0399
(propylene glycol base)

Mixing Valves

Solar Usage Now, Inc. (distributor)
P.O. Box 306
420 E. Tiffin St.
Bascom, OH 44809
(800) 537-0985
(419) 937-2226

(Local plumbing supply stores)

Photovoltaic-Powered Pumps

American Solar King Corp.
7200 Imperial Dr.
P.O. Drawer 7399
Waco, TX 76714–7399
(817) 776-3860

Pumps

Amtrol Inc.
P.O. Box 228
Peru, IN 46970
(317) 472-3351

Edwards Engineering Corp.
101 Alexander Ave.
Pompton Plains, NJ 07444
(201) 835-2808

Grundfos Pumps Corp.
2555 Clovis Ave.
Clovis, CA 93612
(209) 299-9741

Hartell Div. of Milton Roy Co.
70 Industrial Dr.
Ivyland, PA 18974
(215) 322-0730

Hi-Tech, Inc.
3600 16th St.
Zion, IL 60099
(312) 746-2447

ITT Bell & Gossett
4711 Gulf Rd.
Skokie, IL 60076
(312) 677-4030

March Manufacturing, Inc.
1819 Pickwick Ave.
Glenview, IL 60025
(312) 729-5300

Multi-Duti, Inc.
1620 S. Myrtle St.
Monrovia, CA 91016
(213) 357-5091

Myson, Inc.
P.O. Box 5025
Embrey Industrial Park
Falmouth, VA 22403
(703) 371-4331

Paco-Pacific Pumping Co.
P.O. Box 12924
845 92d Ave.
Oakland, CA 94604
(415) 562-5628

Richdel, Inc.
Solar Div.
P.O. Drawer A
Carson City, NV 89701
(702) 882-6786

Taco, Inc.
1160 Cranston St.
Cranston, RI 02920
(401) 942-8000

W. W. Grainger, Inc.
5959 W. Howard St.
Chicago, IL 60648
(312) 647-8900
(Contact Grainger's general offices for local
   distributors)

Storage Tanks (with internal heat exchangers)

Aquus, Inc.
384 Main St.
Norwell, MA 02061
(617) 659-2550

The Electric Heater Co.
P.O. Box 288
45 Seymour St.
Stratford, CT 06497
(203) 378-2659

Ford Products Corp.
Ford Products Rd.
Valley Cottage, NY 10989
(914) 358-8282

Paolino Energy Products, Inc.
726 Kaighns Ave.
Camden, NJ 08103
(609) 541-4614

Solar Products Manufacturing Corp.
1 Alcap Ridge
Cromwell, CT 06416
(203) 635-0266

Solarsystems Industries, Ltd.
#2–11771 Horseshoe Way
Richmond, BC V7A 4S5
Canada
(604) 271-2621

Sunmaster Corp.
35 W. William St.
P.O. Box 1077
Corning, NY 14830
(607) 937-5441

U.S. Solar Corp.
P.O. Drawer K
Hampton, FL 32044
(904) 468-1517

Vaughn Corp.
386 Elm St.
Salisbury, MA 01950
(617) 462-6683

**Thermosiphon System Freeze
  Protection Valves**

Eaton Corp.
Controls Div.
Energy Control Products
191 E. North Ave.
Carol Stream, IL 60187
(312) 260-3055

Chapter 7

**Heat Exchangers for Wood
  and Coal Stoves**

Blazing Showers, Inc.
P.O. Box 327
Point Arena, CA 95468
(707) 882-2592
(stovepipe or firebox heat exchangers)

Cardor Co.
8960 Peninsula Dr.
Traverse City, MI 49684
(616) 941-7426
(thermal flue integral part of solar DHW
  system)

Holly Solar Products, Inc.
P.O. Box 864
Petaluma, CA 94953
(707) 763-6173
(stainless steel or steel mini-tank)

Photic Corp.
2668 S. Memorial Hwy.
Traverse City, MI 49684
(616) 943-9540
(Finductor; externally mounted)

Heat Pump Water Heaters

Airtemp Corp.
Div. of Fedders Corp.
Edison, NJ 08818
(201) 549-7200

Climatrol Sales Co.
Div. of Fedders Corp.
Edison, NJ 08818
(201) 549-7200

Duo-Therm
509 S. Poplar St.
LaGrange, IN 46761
(219) 463-2191

E-Tech
3570 American Dr.
Atlanta, GA 30341
(404) 458-6643

Energy Utilization Systems
201 Seco Rd.
Monroeville, PA 15146
(412) 856-4540

Heat Controller, Inc.
Losey at Wellworth
Jackson, MI 49203
(517) 787-2100

Mor-Flo Industries, Inc.
18450 S. Miles Rd.
Cleveland, OH 44128
(216) 663-7300

Chapter 8

**Complete Solar Systems for Pools,
  Hot Tubs or Spas**

Advanced Energy Products, Inc.
P.O. Box 18485
San Jose, CA 95158
(408) 225-2626
(pool, hot tub or spa)

Applied Solar Products, Inc.
4580–A Almaden Expwy.
San Jose, CA 95118
(408) 269-3050
(pool)

Aquasolar, Inc.
1251 Seeds Ave.
Sarasota, FL 33577
(813) 366-7080
(pool, hot tub or spa)

Bio-Energy Systems, Inc.
221 Canal St.
Ellenville, NY 12428
(914) 647-6700
(SolaRoll pool system)

California Cooperage
880 Industrial Way
San Luis Obispo, CA 93401
(805) 544-9300
(hot tub)

Compool Corp.
1708 Stierlin Rd.
Mountain View, CA 94043
(415) 964-2201
(pool)

Conserdyne Corp.
2444 Wilshire Blvd.
Suite 503
Santa Monica, CA 90403
(213) 829-4326
(pool)

Fafco, Inc.
255 Constitution Dr.
Menlo Park, CA 94025
(415) 321-3650
(pool, hot tub or spa)

Paolino Energy Products, Inc.
726 Kaighns Ave.
Camden, NJ 08103
(609) 541-4614
(pool/DHW, hot tub/DHW and pool or hot
 tub)

Purex Pool Products, Inc.
18400 E. Mohr Ave.
City of Industry, CA 91749
(213) 965-1551
(pool, hot tub or spa)

Raypak, Inc.
31111 Agoura Rd.
Westlake Village, CA 91361
(213) 889-1500
(pool, hot tub or spa)

Sealed Air Corp.
Park 80 Plaza E.
Saddle Brook, NJ 07662
(800) 631-3818
(800) 562-2728 (in New Jersey)
(201) 797-4000
(pool, hot tub or spa)

Solar Development, Inc.
3630 Reese Ave.
Riviera Beach, FL 33404
(305) 842-8935
(pool, hot tub or spa)

Solar Industries, Inc.
2300 Hwy. 34
Manasquan, NJ 08736
(201) 223-8100
(pool or spa)

Solar Living, Inc.
P.O. Box 12
Netcong, NJ 07857
(201) 691-8483
(pool and hot tub)

Solar Research Systems, Inc.
2116 S. Yale St.
Santa Ana, CA 92704
(714) 540-4292
(pool, spa or hot tub)

Sunglo Solar, Ltd.
1081 Alness St.
Downsview, ON M3J 2J1
Canada
(416) 661-2560
(pool)

Differential Controllers for Pools, Hot Tubs and Spas

Heliotrope General
3733 Kenora Dr.
Spring Valley, CA 92077
(714) 460-3930
(pool, hot tub or spa)

Raypak, Inc.
31111 Agoura Rd.
Westlake Village, CA 91361
(213) 889-1500
(pool, hot tub or spa)

Rho Sigma, Inc.
Sub. of Watsco, Inc.
1800 W. 4th Ave.
Hialeah, FL 33010
(305) 885-1911

Richdel, Inc.
Solar Div.
P.O. Drawer A
Carson City, NV 89701
(702) 882-6786
(pool, hot tub or spa)

Differential Controllers for Pools, Hot Tubs and Spas — *continued*

Solar Components Corp. (distributor)
P.O. Box 237
Manchester, NH 03105
(603) 668-8186
(pool, hot tub or spa)

Solar Industries, Inc.
2300 Hwy. 34
Manasquan, NJ 08736
(201) 223-8100
(pool)

Solar Usage Now, Inc. (distributor)
P.O. Box 306
420 E. Tiffin St.
Bascom, OH 44809
(800) 537-0985
(419) 937-2226
(pool, hot tub or spa)

Pool Collectors (and their absorber material)

Ameropean Corp.
71 Hartford Tpk. S.
Wallingford, CT 06492
(203) 265-4648
(polyethylene)

Applied Solar Products, Inc.
4580–A Almaden Expwy.
San Jose, CA 95118
(408) 269-3050
(PVC)

Aquasolar, Inc.
1251 Seeds Ave.
Sarasota, FL 33577
(813) 366-7080
(ABS pipe)

Bio-Energy Systems, Inc.
221 Canal St.
Ellenville, NY 12428
(914) 647-6700
(SolaRoll)

Fafco, Inc.
255 Constitution Dr.
Menlo Park, CA 94025
(415) 321-3650
(polypropylene)

Grumman Allied Industries, Inc.
445 Broad Hollow Rd.
Plant 9, Floor #4
Melville, NY 11747
(516) 454-8600
(copper tubes and aluminum fins)

Miroson Plastic Solar Products Inc.
P.O. Box 14
Mockingbird La.
RR 1
Camlachie, ON N0N 1E0
Canada
(519) 869-4313
(Rovel by Uniroyal)

Mor-Flo Industries, Inc.
18450 S. Miles Rd.
Cleveland, OH 44128
(216) 663-7300
(copper)

Ottawa Solartronics, Ltd.
1070 Morrison Dr.
Ottawa, ON K2H 8K7
Canada
(613) 820-3274
(Noryl; polyphenelyne oxide)

Pure Pool Products, Inc.
18400 E. Mohr Ave.
City of Industry, CA 91749
(213) 965-1551
(copper)

Raypak, Inc.
31111 Agoura Rd.
Westlake Village, CA 91361
(213) 889-1500
(copper or copper tubes with aluminum
fins)

Sealed Air Corp.
Park 80 Plaza E.
Saddle Brook, NJ 07662
(800) 631-3818
(800) 562-2728 (New Jersey)
(ethylene-propylene copolymer)

Solar Development, Inc.
3630 Reese Ave.
Riviera Beach, FL 33404
(305) 842-8935
(SolaRoll)

Solar Industries, Inc.
2300 Hwy. 34
Manasquan, NJ 08736
(201) 223-8100
(polypropylene)

Solar Oriented Environmental Systems, Inc.
10639 S.W. 185 Terrace
Miami, FL 33157
(305) 233-0711
(polybutylene in cement)

Solar Research Systems, Inc.
2116 S. Yale St.
Santa Ana, CA 92704
(714) 540-4292
(polypropylene)

Sunglo Solar, Ltd.
1081 Alness St.
Downsview, ON M3J 2J1
Canada
(416) 661-2560
(copolymer polypropylene)

Sun-West Solar Systems, Inc.
4024 E. Broadway Rd.
Suite 1001
Phoenix, AZ 85040
(602) 243-6171
(polyethylene/ABS headers)

Sunworks, Inc.
P.O. Box 3900
Somerville, NJ 08876–1270
(201) 469-0399
(polypropylene)

Valves for Pool, Hot Tub or Spa Systems

Compool Corp.
1708 Stierlin Rd.
Mountain View, CA 94043
(415) 964-2201
(automatic diverter)

Ortega Valve and Engineering Co.
Div. of Purex Pool Products, Inc.
14902 Moran St.
Westminster, CA 92683
(714) 892-0102
(manual or automatic diverter)

Richdel, Inc.
Solar Div.
P.O. Drawer A
Carson City, NV 89701
(702) 882-6786
(automatic diverter and gate)

Solar Usage Now, Inc. (distributor)
P.O. Box 306
420 E. Tiffin St.
Bascom, OH 44809
(800) 537-0985
(419) 937-2226
(manual or automatic gate)

Chapter 9

Btu Meters

Watsco, Inc.
1800 W. 4th Ave.
Hialeah, FL 33010
(305) 885-1911

Flow Meters

Blue White Industries, Inc.
14931 Chestnut St.
Westminster, CA 92683
(714) 893-8529

Solar Usage Now, Inc. (distributor)
P.O. Box 306
420 E. Tiffin St.

Bascom, OH 44809
(800) 537-0985
(419) 937-2226

Taco, Inc.
1160 Cranston St.
Cranston, RI 02920
(401) 942-8000

APPENDIX 2
SOLAR ENERGY ASSOCIATIONS

American Solar Energy Society (ASES)

American Solar Energy Society
1230 Grandview Ave.
Boulder, CO 80302
(303) 492-6017

Local Chapters of ASES

Alabama Solar Energy Association
c/o Johnson Environmental and Energy
  Center
University of Alabama
P.O. Box 1247
Huntsville, AL 35889

Arizona Solar Energy Association
P.O. Box 25396
Phoenix, AZ 85002

Colorado Solar Energy Association
P.O. Box 1284
Alamosa, CO 81101

Eastern New York Solar Energy Association
P.O. Box 5181
Albany, NY 12205

Florida Solar Energy Association
P.O. Box 248271
University Station
Coral Gables, FL 33124

Georgia Solar Energy Association
P.O. Box 32748
Atlanta, GA 30332

Illinois Solar Energy Association
P.O. Box 1592
Aurora, IL 60507

Iowa Solar Energy Association
P.O. Box 68
Iowa City, IA 52244

Kansas Solar Energy Society
P.O. Box 8516
Wichita, KS 67208

Metropolitan New York Solar
  Energy Society
P.O. Box 2147
Grand Central Station
New York, NY 10163

Michigan Solar Energy Association
696 N. Mill St., #100
Plymouth, MI 48170

Mid-Atlantic Solar Energy Association
2233 Gray's Ferry Ave.
Philadelphia, PA 19146

Minnesota Solar Energy Association
P.O. Box 762
Minneapolis, MN 55440

Mississippi Solar Energy Association
225 W. Lampkin Rd.
Starkville, MS 39759

Nebraska Solar Energy Association
c/o Prof. John Thorp
University of Nebraska at Omaha
60th and Dodge Sts.
Engineering Building 110
Omaha, NE 68182

Nevada Solar Energy Advocates
P.O. Box 8179
University Station
Reno, NV 89507

New England Solar Energy Association
P.O. Box 541
Brattleboro, VT 05301

New Mexico Solar Energy Association
P.O. Box 2004
Santa Fe, NM 87501

North Carolina Solar Energy Association
P.O. Box 10431
Raleigh, NC 27605

Northern California Solar Energy
  Association
P.O. Box 886
Berkeley, CA 94701

Ohio Solar Energy Association
c/o Columbus Technical Institute
550 E. Spring St.
Columbus, OH 43216

Solar Energy Association of Oregon
2637 S.W. Water
Portland, OR 97201

South Dakota Renewable Energy
  Association
P.O. Box 782
Pierre, SD 57501

Tennessee Solar Energy Association
c/o Eric Lewis
4500 Idaho Ave.
Nashville, TN 37219

Texas Solar Energy Society
600 W. 28th, Suite 101
Austin, TX 78705

Local Chapters of ASES — *continued*

Virginia Solar Energy Association
c/o Piedmont Technical Associates
300 Lansing Ave.
Lynchburg, VA 24503

Washington Solar Council
1932 1st Ave., Suite 917
Seattle, WA 98101

Solar Energy Resource Association
of Wisconsin
Division of State Energy
P.O. Box 7868
Madison, WI 53709

FURTHER READING

Books

American Society of Heating, Refrigerating, and Air-Conditioning Engineers (ASHRAE). *Methods of Testing to Determine the Thermal Performance of Solar Collectors.* New York: ASHRAE, 1978. (Available from the ASHRAE Publications Sales Department, 1791 Tullie Circle, NE, Atlanta, GA 30329. Order no. 93–77.)

––––––. *Methods of Testing to Determine the Thermal Performance of Solar Domestic Water Heating Systems.* Atlanta: ASHRAE, 1981. (Available from the ASHRAE Publications Sales Department, 1791 Tullie Circle, NE, Atlanta, GA 30329. Order no. 95–1981.)

Bainbridge, David A. *The Integral Passive Solar Water Heater Book.* Davis, Calif.: The Passive Solar Institute, 1981.

Barada, William R. *Build Your Own Solar Water Heater.* Winter Park, Fla.: Florida Conservation Foundation, 1976.

Bryenton, Roger; Cooper, Ken; and Mattock, Chris. *The Solar Water Heater Book.* Toronto: Renewable Energy in Canada, 1980.

Butti, Ken, and Perlin, John. *A Golden Thread.* Palo Alto, Calif.: Cheshire Books, 1980.

California Energy Commission. *Directory of Certified Water Heaters*. Sacramento: California Energy Commission, 1980. Available from the Publications Department, California Energy Commission, 1111 Howe Ave., Sacramento, CA 95825. Order no. 400–00–020.

————. *Test Results from Testing and Inspection Program for Solar Equipment*. Sacramento: California Energy Commission, 1980. (Available from the Publications Department, California Energy Commission, 1111 Howe Ave., Sacramento, CA 95825. Order no. P500–80–056.)

Campbell, Stu, and Taff, Doug. *Build Your Own Solar Water Heater*. Charlotte, Vt.: Garden Way Publishing, 1978.

Carter, Joe, ed. *Solarizing Your Present Home*. Emmaus, Pa.: Rodale Press, 1981.

Dean, Thomas Scott, and Hedden, Jay W. *How to Solarize Your Home*. New York: Charles Scribner's Sons, 1980.

Ecotope Group. *A Solar Water Heater Workshop Manual*. Seattle, Wash.: Ecotope Group, 1979.

Franklin Research Center. *Installation Guidelines for Solar DHW Systems in One- and Two-Family Dwellings*. Washington, D.C.: U.S. Government Printing Office, 1979. Order no. 023–00–00520–4.

Fried, Chris. *SunShuttle Solar Water Heater: Assembly and Installation Manual*. Catawissa, Pa.: Chris Fried Solar, 1982. (Available for $6.95 from Chris Fried Solar, R.D. 3, Box 229G, Catawissa, PA 17820.)

Hollibaugh, Bill. *The Holly Hydro Heater Owner's Handbook*. Petaluma, Calif.: Holly Solar Products, Inc, 1982. (Available for $2.50 from Holly Solar Products, Inc., P.O. Box 864, Petaluma, CA 94953.)

Kirby, James, and Mirvis, Kenneth, eds. *Solar Energy: An Installer's Guide to Domestic Hot Water*. Washington, D.C.: National Association of Solar Contractors, 1982.

Klima, Jon. *The Solar Controls Book: Fundamentals of Domestic Hot Water and Space Heating Solar Controls*. Lakewood, Colo.: Solar Training Publications, 1982.

McPherson, Beth. *Hot Water from the Sun*. Washington, D.C.: U.S. Government Printing Office, 1981. Order no. HUD–PDR–548.

Milne, Murray. *Residential Water Re-Use*. Davis, Calif.: California Water Resources Center, University of California at Davis, 1979. (Available for $10 from California Water Resources Center, University of California, Davis, CA 95616.)

————. *Residential Water Conservation*. Davis, Calif.: California Water Resources Center, University of California at Davis, 1976. (Available from the National Technical Information Service, 5285 Port Royal Rd., Springfield, VA 22161. Order no. PB253253.)

Books — *continued*

Montgomery, Richard H., and Budnick, James. *Solar Decision Book: Your Guide to Making a Sound Investment.* New York: John Wiley & Sons, 1978.

New York State Energy Research and Development Authority (NYSERDA). *Consumer's Guide to Buying Solar Domestic Hot Water.* Albany, N.Y.: NYSERDA, 1982. (Available from NYSERDA, Solar Publications, Two Rockefeller Plaza, Albany, NY 12223.)

Palla, Robert L., Jr. *The Potential for Energy Savings with Water Conservation.* Washington, D.C.: U.S. Department of Commerce, 1979. (Available from the National Technical Information Service, 5285 Port Royal Rd., Springfield, VA 22161. Order no. PB80127202.)

Root, Douglass E., Jr., and Partington, William M., Jr. *Solar Heating for Swimming Pools.* Winter Park, Fla.: Florida Conservation Foundation, 1980.

Sharpe, William E., and Fletcher, Peter W., eds. *Proceedings: Conference on Water Conservation and Sewage Flow Reduction with Water-Saving Devices.* State College, Pa.: Pennsylvania State University, 1975.

Shaw, Nadine, and Bauer, Janice. *A New Jerseyan's Consumer Guide to Solar Energy Systems.* Trenton, N.J.: New Jersey Public Interest Research Group, 1978.

Shepherd, P. B. *Performance Evaluation of Point-of-Use Water Heaters.* Washington, D.C.: U.S. Department of Commerce, 1980. (Available from the National Technical Information Service, 5285 Port Royal Rd., Springfield, VA 22161. Order no. ADA091843.)

Sigworth, Harry W., Jr. "Keeping Swimming Pools Warm." Unpublished monograph, 1981.

Smith, Robert O., and Associates. *Report Summary of Performance Problems of 100 Residential Solar Water Heaters Installed by New England Electric Company Subsidiaries in 1976 and 1977.* Newton Highlands, Mass.: Robert O. Smith and Associates, 1977. (Available for $11 from Robert O. Smith and Associates, 55 Chester St., Newton Highlands, MA 02161.)

Solar Age magazine. *Solar Products Specifications Guide.* Harrisville, N.H.: SolarVision, 1982.

Solar Environmental Engineering Co., Inc. *Solar Domestic Hot Water System Inspection and Performance Evaluation Handbook.* Washington, D.C.: Department of Energy, 1981. (Available from the National Technical Information Service, 5285 Port Royal Rd., Springfield, VA 22161. Order no. DE82006184.)

Sussman, Art, and Frazier, Richard. *Handmade Hot Water Systems.* Point Arena, Calif.: Garcia River Press, 1978.

Zweig, Peter. *Inexpensive Do-It-Yourself Solar Water Heater or Pre-Heater.* Occidental, Calif.: Farallones Institute, 1978.

Magazines

National Energy Journal, 411 Cedar Rd., Chesapeake, VA 23320.

New Shelter, 33 E. Minor St., Emmaus, PA 18049.

Popular Science, 380 Madison Ave. New York, NY 10017.

Solar Age, P.O. Box 985, Farmingdale, NY 11737.

Solar Engineering and Contracting, P.O. Box 3600, 755 W. Big Beaver Rd., Troy, MI 48099.

Solar Heating and Cooling, Gordon Publications, P.O. Box 360, Dover, NJ 07801.

Sun-Up, P.O. Drawer S, Yucca Valley, CA 92284.

INDEX

Page numbers in **boldface** type indicate tables.